Innovations in Automotive and Aerospace Assembly

R-453 Related Resources

Manufacturing System and Process Development for Vehicle Assembly
He Tang
(PRODUCT CODE: R-457)

Automotive Vehicle Assembly Processes and Operations Management
He Tang
(PRODUCT CODE: R-456)

Automated/Mechanized Drilling and Countersinking of Airframes
George Nicholas Bullen
(PRODUCT CODE: R-416)

Composite Materials Handbook
(PRODUCT CODE: R-422.SET6)

Lightweight Materials: Understanding the Basics
F.C. Campbell
(PRODUCT CODE: B-ASM-041)

Structural Composite Materials
F.C. Campbell
(PRODUCT CODE: B-ASM-031)

Fatigue and Durability of Structural Materials
G.R. Halford, S.S. Manson
(PRODUCT CODE: B-ASM-009)

Additive Manufacturing of Aerospace Composite Structures: Fabrication and Reliability
Rani Elhajjar
(PRODUCT CODE: PT-181)

Aluminum Auto-Body Joining
George Nicholas Bullen
(PRODUCT CODE: PT-173)

Innovations in Automotive and Aerospace Assembly

GEORGE N. BULLEN

SAE INTERNATIONAL

Warrendale, Pennsylvania, USA

400 Commonwealth Drive
Warrendale, PA 15096-0001 USA
E-mail: CustomerService@sae.org
Phone: 877-606-7323 (inside USA and Canada)
724-776-4970 (outside USA)

Library of Congress Catalog Number 2015957326
SAE Order Number R-453
http://dx.doi.org/10.4271/R-453

R-453 Innovations in Automotive and Aerospace Assembly

ISBN-Print 978-0-7680-8281-4
ISBN-HTML 978-0-7680-9164-9
ISBN-epub 978-0-7680-8468-9
ISBN-prc 978-0-7680-8467-2

To purchase bulk quantities, please contact: SAE Customer Service

E-mail: CustomerService@sae.org
Phone: 877-606-7323 *(inside USA and Canada)*
724-776-4970 *(outside USA)*

Visit the SAE International Bookstore at books.sae.org

dedication

To the mechanics that assemble the airplane and the automobile—
two of the world's most complex machines.
You continue to keep me grounded and add clarity to the nuances and challenges of
precisely assembling dissimilar materials into progressively larger complex shapes.

Also
To my mother, who taught me to dream;
Dr. Hall (Leuzinger High School), who taught me to communicate my dreams;
Dr. Hall (Ranchvale Elementary School), who taught me that any complex puzzle can
be understood and solved by breaking it down to its component pieces.

contents

CHAPTER 6
Summary 83

Nondestructive Adhesive Bond-Line Inspection

CHAPTER 7
Introduction 99

preface

This book has been written to identify and describe two emerging disruptive manufacturing process technologies that will significantly impact the way airplanes and automobiles are assembled.

The book is divided into two sections. Both sections address the complexities of assembling the myriad of parts made from dissimilar materials comprising an airplane, automobile, or space vehicle. Both sections also identify technical solutions that will positively impact the assembly process when parts made from composite or aluminum materials are joined into an airplane or automotive assembly. The technologies will impact assembly processes by reducing assembly cost, improving throughput, and providing the diagnostic data and analysis to empower the use of advanced analytics.

The airplane and automobile assembly processes are complicated by use of composite materials. Using parts made from composite materials either joined with other composite parts, or when composite parts are joined to metal parts introduces complexity and variability into the assembly process. The automotive industry has also introduced and begun integrating composites and high-grade aerospace aluminum onto the structural components of the road vehicles they manufacture. Joining parts made from composites or aluminum to dissimilar materials requires processes and controls that ensure the dimensional and structural integrity of the assembly once it is joined.

Section One addresses airplane and automotive assemblies that use fasteners as their primary means to join parts together and describes a revolutionary technology that has emerged to significantly enhance quality and reduce cost.

Section Two describes the use of adhesives as the primary means to join and hold parts together, and the emerging technologies that will ensure joining of parts are robust and meet design specifications without affecting the bond-line adhesion properties.

The first section addresses the technical solution for the process of drilling-and-filling holes with fasteners. Fasteners are and will continue to be the primary way to hold airplane parts together in the foreseeable future. Fasteners are also addressed in **Section One** for automotive assembly due to the increased migration of composites and aluminum onto automotive structural components. Much of the focus since composite materials have made their way onto the airplane and automobile has been on fabrication of parts to be later assembled. It continues to be a primary focus for conferences and technology development. At the recent International Composites Conference on Materials (ICCM), of the 1500 papers presented, only 7 were about the assembly process.

This is surprising given that 65% of the cost of an airframe is in the assembly process, and 65% of the cost, 80% of the defects, and 80% of all lost time injuries occur during assembly operations, including the drill-and-fill process. Automobiles and transportation vehicles contain the same cost and quality components during their assembly processes. There is evidence that the formerly neglected composite-part drill-and-fill process has recently been recognized as a cost and quality driver, especially in aerospace.

The assumption that the same hole-bore evaluation methods and gauges that were formerly used on metals can be transferred to composite hole-bore evaluation is wrong. The use of gauges specifically designed for metals results in defects because of the complex and variable nature inherent in composite materials. Inconsistent thickness and internal material variability of cured composite material complicate the drilling and hole-bore quality evaluation process. Each hole drilled and countersunk in composites, or through composites and into metal reacts differently no matter how controlled and stable the process. It has become a process requirement that each hole be inspected by gauge and human "eyes-on" evaluation. The insertion of a manual operation into the drill-and-fill process has frustrated the application of a fully automated process in aerospace and is a complex process-challenge emerging in the automotive industry.

Therefore, there has been a concerted effort to identify and develop a technical solution to evaluate all the parameters for composite hole-bore acceptability with a single gauge. And use the gauge to eliminate the current process requirement to have humans visually inspect 100% of the holes that are drilled and countersunk. Eliminating the number of gauges, increasing the fidelity of hole evaluation (specific to composites), and removing the des-tack[1] requirement with a single technical solution would significantly reduce cost, increase quality, and enable the complete drill-and-fill automated solution for assembling the airplane and, by extension, automobiles.

The first section describes the latest technology that provides that solution.

Section One describes a technology solution to accelerate the rate for installation of fasteners at a higher level of quality with fewer process steps. **Section Two** is written to reflect the desire to eliminate or significantly reduce fasteners. The development of a technical solution to facilitate the elimination or reduction in fastener count would solve the fastener process-complexity issue. One natural benefit of composites (and to some extent aluminum) is their ability to be adhesively bonded together rather than fastened. One approach is to "fix" the complex, costly, and frustrating process of drilling and countersinking holes to simply eliminate as many fasteners as possible by combining multiple pieces into a unified structure or adhesively bonding the pieces together.

There are many research projects that strive to accomplish an adhesively bonded or mostly bonded airplane or automobile. The inhibitor to the adhesively bonded-assembly effort is the ability to confirm that the hidden bond-line meets the necessary strength properties. Once bonded together, the bond-line is hidden from sight and destructive testing is the only positive way to validate that a robust bond-line exists. Of course, once the parts are destroyed they cannot be used.

The other method is the "trust-me" method for adhesive bonding. This process employs Draconian[2] manufacturing process controls that are applied so there is a high degree of confidence that the bond-line is robust. The process relies on upstream manufacturing control to ensure a robust bond-line without validation once the assembly is completed. One research program is called TRUST. Lockheed Martin Aeronautics Co.

---

[1] De-stack describes a process when parts are disassembled after drilling to clean, de-burr, and inspect the parting plane between the two parts.

[2] A code of laws prepared by Draco the celebrated lawgiver of Athens that by modern standards are considered exceedingly severe. The term draconian has come to be used to refer to any unusually harsh law.

(LM Aero Palmdale) lead the Transition Reliable Unitized STructure (TRUST) project, part of the Defense Advanced Research Projects Agency's (DARPA, WA, DC, and USA) Open Manufacturing program, an effort to reduce cost and speed delivery of high-quality parts with predictable performance for defense systems.

Customers and manufacturing experts are reticent to employ a "trust" manufacturing process, because experience has repeatedly demonstrated that no matter how rigid and "controlled" the manufacturing process is, manufacturing escape defects[3] occur. It is a noble goal to push toward a zero-defects product. The risk of applying a "trust" vision for zero defects that relies on repeatedly assuming each assembled vehicle has passed through the highly complex manufacturing process without an escape defect is amplified by the fact that most vehicles carry people. And the people that the vehicle carries have a reasonable expectation that the vehicle they are riding on is safe and reliable.

The reluctance of manufacturers to employ a "trust" manufacturing method is also driven by cost. The more detailed the oversight and controls placed on the manufacturing process, the higher the manufacturing cost.

Neither the "trust" method nor the destroy-to-confirm method is an optimal solution. Therefore, vigorous research initiatives have focused on developing and deploying technologies that can "see" into the adhesively joined parts to evaluate the adhesive bond-line for robustness and conformance to specification.

Section Two describes these technology development efforts and the resultant technologies, their technology maturity levels, and technology approach. When applied, they can potentially provide the necessary adhesive bond-line conformance validation for either defect discovery or adhesive bond-line conformance to specification. And offer to provide an adhesive bond-line confirmation technology that will be automated and effectively integrated into the manufacturing process without significant increase to the complexity of the manufacturing process.

Section Two also provides a repair process to fix areas of the bond-line that the newly developed technology identifies as deficient. Adhesive bond-line inspection technologies described in **Section Two** primarily focus on either validation or defect discovery. The natural outcome of any defect detection/discovery is, "Now what?" Any gauge or evaluation process must be accompanied by a repair procedure. Repair procedures are described as a component of **Section Two**.

Digital technology maturation has enabled the genesis of technologies to capture part and assembly DNA.[4] As parts and assemblies are manufactured, they begin to take on unique attributes that differentiate them from other parts and assemblies that are identified as the same. Capturing and storing the critical history of defects, repairs, and variations of each part or assembly is critical to sustainability, maintenance, and actuarial science for vehicle life cycle and performance prediction.[5] **Section Two 2** will contain a description of the three components for manufacturing visibility: part/assembly

---

[3] An escape defect is a defect that remains undetected during the manufacturing process. Critical escape defects can affect product performance or cause vehicle operational failure that in some cases is catastrophic.

[4] The fundamental and distinctive characteristics or qualities of someone or something, especially when regarded as unchangeable.

[5] Actuarial science is the discipline that applies mathematical and statistical methods to assess risk.

adhesive bond-line evaluation and defect detection, defect repair processes if a defect is discovered, and technology that can be employed to embed the part health and history into the part or assembly.

Both sections of this book provide a description of the functional operations of the respective technologies. They also provide the necessary cost–benefits analysis data to substantiate the process cost saving derived from fielding and integrating these technologies into the airplane or automotive assembly process.

The disruptive technologies described in this book and the manufacturing processes and products they enhance are critical to the evolution for the assembly of a modern airplane, automobile, or other transportation vehicle such as high speed rail-type trains. They are essential components to control cost, improve quality, and enhance sustainability through the products' life cycle. However, their most important contribution is their ability to fully integrate the benefits of composites and dissimilar material into the assembly process to make safer and more cost-effective airplanes, automobiles, and transportation vehicles. Regardless of the evolution of bonded structures, fasteners will continue to be with us for a long time. The integration of the two technologies into assembly operations will eventually morph into a future airplane/automotive/transport vehicle assembly where a minimum number of fasteners will be strategically placed to complement the bonded structure.

The disruptive technologies described in this book are critical components and complement the vision for a *Fully Optimized Digitally Enabled Intelligent Factory Enterprise* to manufacture airplanes, space, automobiles, and transportation vehicles.

Acknowledgments

I would like to acknowledge SAE International for the opportunity to identify and describe the disruptive technologies that will change the way airplanes and automobiles are assembled.

Also, for their innovation and investment for the progress, understanding, development, and maturation of advanced assembly technologies for application on airplanes, automobiles, and space vehicles.

Air Force Research Labs (AFRL/RX)—Materials and Manufacturing
Wright-Patterson Air Force Base,
Dayton, OH 45424, USA

Airbus
Research and Technology,
AIRBUS Operations, S.L.,
John Lenon s/n, 28906 Madrid, Spain

Bell Helicopter Textron Inc
600 E Hurst Blvd,
Hurst, TX 76053, USA

Boeing Airplane Company
Paine Field,
Everett, WA, USA

Comau
21175 Telegraph Rd.,
Southfield, MI 48033, USA

Defense Advanced Research Projects Agency (DARPA)
675 N Randolph St.
Arlington, VA 22203, USA

Douglas O'Neil
Silver Strand Beach,
Oxnard, CA 93035, USA

Electro Impact
4413 Chennault Beach Rd.,
Mukilteo, WA 98275, USA

Kuka Robotics Corporation, Aerospace
51870 Shelby Parkway,
Shelby Township, MI 48315, USA

Lockheed Martin
1 Lockheed Blvd,
Fort Worth, TX 76108, USA

LSP Technologies
6145 Scherers Place,
Dublin, OH 43016, USA

M Torres
Ctra. Pamplona-Huesca, Km 9,
Torres de Elorz, 31119 Navarra, Spain

NAVAIR—US Navy—Composites Laboratory
Point Mugu, CA 93042, USA

Naval Research Labs (NRL)
875 N Randolph St.
Arlington, VA 22217, USA

Northrop Grumman Corporation
825 N Douglas St.
El Segundo, CA 90245, USA

The United Sciences Team
2277 Peachtree Road Northeast, Suite D,
Atlanta, GA 30309, USA

Section One

Noncontact Ring-Laser Metrology Systems

CHAPTER 1

Introduction

Any sufficiently advanced technology is equivalent to magic.

—**Sir Arthur C. Clarke**

1.1 Introduction

It is important to understand how a process evolved to its current state of maturity to fully understand the impact of a new disruptive technology, and how it will revolutionize an industry. History leads us to our present state and empowers our vision for the future.

Material types have driven manufacturing to adapt and align the fabrication and assembly processes to facilitate a quality product that is reliable, can be manufactured at a price point that is affordable, and be manufactured in quantities that can be widely distributed. Up until the past two decades, aluminum in airplanes and steel in automobiles were the primary materials used to produce these two complex machines. Metal-to-metal assemblies, specifically the same type of metal-to-metal assemblies, have distinct manufacturing process advantages over dissimilar metal assemblies, or more recently, the highly complex manufacturing processes required for composite materials over metal assemblies.

1.2 Metals

Metals used in the construction of airplanes and other transportation vehicles have well-understood properties. Metal pieces and parts also have manufacturing processes that were developed and specifically aligned to facilitate assemblies that are joined to meet defined specifications for form, fit, and function. Gauges and measurement instruments are used along the manufacturing trail to ensure that parts, pieces, and assemblies aggregate up into the desired product that conforms to engineering specifications.

Metals are also isotropic. In the study of the mechanical properties of materials, "isotropic" means having identical values of a property in all directions. Glass and metals are examples of isotropic materials. Working with materials having isotropic properties is valuable in manufacturing for many reasons. Isotropic materials when fabricated and joined are predictable and uniform. Technology, experience, and talent render today's raw metal materials into forms that are close to their finished shape and size. Little or, in many cases, no work needs be done to finish the parts. Removing a minimum amount of material or eliminating post-process straightening reduces stresses created from machining, grinding, drilling, and finishing operations. Sheet metal, especially aluminum, comes in uniform thicknesses that vary as little as 0.0001 in. (0.00254 mm) with surface finish and flatness that facilitate a precise parting plane surface between two joined pieces of material.

Decades of working with metals, particularly aluminum and steel, has led to a deep understanding of their properties. The evolution of computer-aided manufacturing has advanced the ability of manufacturing to reduce variability. Families of metal parts are joined with little or no gap between a sheet metal covering piece and a machined structural substructure. This close fit-up attribute streamlines production for many reasons. A tight fit reduces corrosion and debris migration between parts, facilitates a tight rub-free joining, and reduces post-process assembly activity. **Figure 1.1** shows how two pieces of metal material fit tightly together to create a pristine mating plane surface for assembled parts.

Having one part with a surface that is tightly fit to another when combined with precise thickness control streamlines assembly by reducing variability. Two ways that a predictably precise thickness of two parts facilitates streamlined assembly efficiency are fastener installation and welding. By having a highly predictable dimension from the front side to the backside and with a tight surface fit between two pieces of material, a welding machine can produce a more consistent weld. And the grip length for fasteners need not be measured after holes are drilled to validate the right fastener length. Grip length is the total distance between the underside of the nut to the bearing face of the bolt head, including washer, gasket thickness, and so on, so that fasteners provide the necessary "grip" to hold the parts together to meet design specified requirements. Therefore, mechanics or machines can weld or install fasteners without measuring every point where a joining method is applied. The characteristics of uniform thickness and tight surface fit have the added advantage of enabling automation for these highly repetitive processes. **Figure 1.2** shows the advantage of a uniform thickness for fastener grip length installation consistency. Uniform thickness has the added advantage of reducing the length selection of fasteners that must be stocked.

FIGURE 1.1 Metal-to-metal tightly fitting mating surface at assembly.

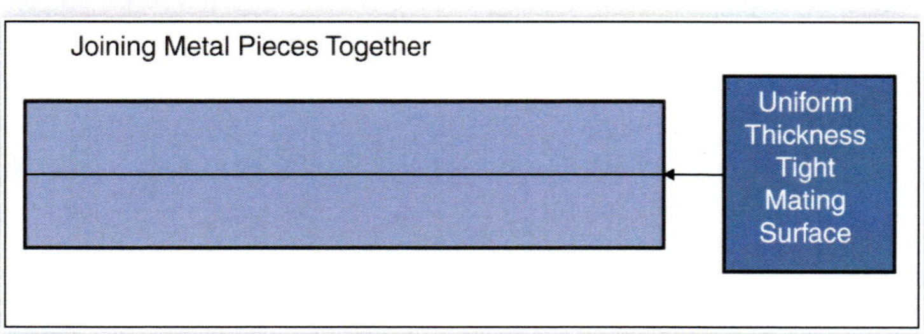

While parts are machined and sheet material is produced to exacting tolerances, the forming of sheet metal material can vary slightly to the desired shape. At assembly fit-up a sheet metal part may have spring-back or spring-in that prohibits the sheet metal part from lying down precisely against the machined part, or against another sheet metal part. Metal can be bent and clamped into position to mitigate this condition and thus bring the part into a tight marriage with the surface of its mating part. And it can be reasonably "bent" into a conforming position without creating unacceptable stresses that reduce strength. Of course, analysis should be done to correct spring-in or spring-out to attain the best fit if it exceeds the manufacturing allowable specified limits or slows the assembly. The point here is that metal is a forgiving material that can be adjusted at assembly if needed.

FIGURE 1.2 Fastener grip length uniformity provided by metal assemblies.

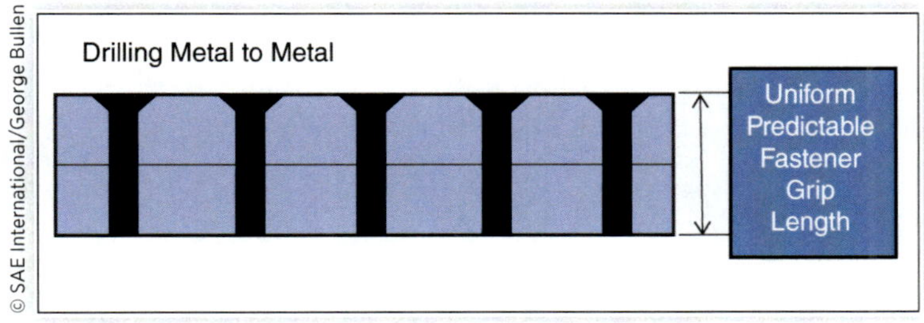

Metals are also very "machine friendly." Tools such as computer models, finite element analysis, numerical controlled machines, and decades of experience have led to a precise understanding of how metals react to machining, grinding, drilling, and finishing work. A large array of gauges has been developed that assume the predictability of metal parts. They measure the specific and critical attributes of parts and assemblies to ensure that they meet the dimensions identified by design engineering necessary for safe and reliable operation of the product. Metal provides a level of repeatability and consistency that instills sufficient confidence where conformance assumptions can be made that accelerate the manufacturing process without compromising the quality of the product. These include random inspection rhythms at the macro level, where a robust and proven consistent manufacturing process reduces inspection intervals for specific attributes and even parts. And at the micro level where reasonable assumptions are made, given the attributes of metal. An example of macro- and micro-level assumptions for gauging and inspection of metal is resident in aerospace fastener holes.

When two pieces of metal are joined by means of a fastener, a hole must be drilled to a specified diameter to facilitate the insertion of a proper sized fastener. The hole must be drilled larger than the fastener and be concentric. In most cases, the fastener hole must be countersunk to accept a flush fastener. The primary inspection criterion is to gauge the hole-bore diameter, countersink depth, and grip length of the material stack. This requires in most cases, three gauges.

At the macro level, as the production line matures, the interval for inspection of these three attributes is reduced from 100% of the holes to 10%, and in some areas below 10%. However, it is critical for inspection to take place to validate the conformance of these three attributes to ensure the fastener provides the level of fidelity needed for holding the assembly together. Even if the properties of the metal are well understood, the relationship between the hole-bore size and concentricity, the countersink depth, and the grip length to the fastener are so critical to the strength and performance of the assembly, they must be inspected to ensure the relationship is sustained.

At the micro level, the attributes are measured relying on the known properties and attributes of the metals to make assumptions and projections. Gauges have evolved to measure metal. **Figure 1.3** shows a countersink gauge where the cone of the gauge touches

FIGURE 1.3 Countersink gauge used to determine countersink depth conformance for flush fasteners.

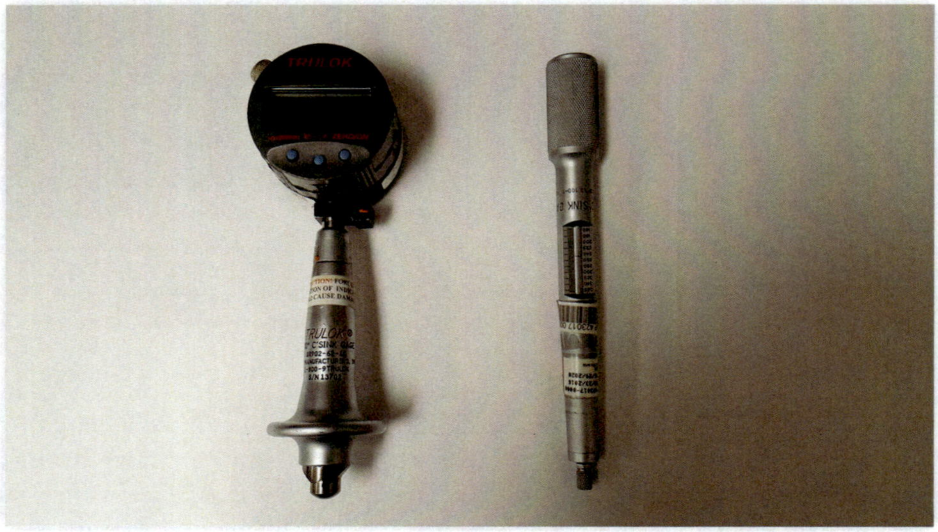

© SAE International/George Bullen

the countersink and then is depressed to gauge the depth. The assumption is that the manufacturing process has produced a uniform and round cone-shaped countersink. And that the metal material has been removed uniformly. All the gauge is measuring is depth and not the quality of the countersink surface where metal has been removed. Metal removal processes and metal material are reliably predictable enough for an assumption of conformance to be made.

Figure 1.4 shows the operation of a typical hole-bore inspection gauge process. The gauge operates using a shaft that is close to the inside diameter of the hole it is to inspect. Small balls in the shaft roll against the inside of the hole-bore as it is inserted. The balls are located at 180° from one another. Compression of the balls sends a signal to the gauge software that measures the bore diameter at their points of contact as the gauge is inserted along a line from the front to the back of the hole-bore. After the gauge exits the backside of the hole-bore, the gauge is turned 90° and retracted, measuring lines

FIGURE 1.4 Hole-bore operation assumptions.

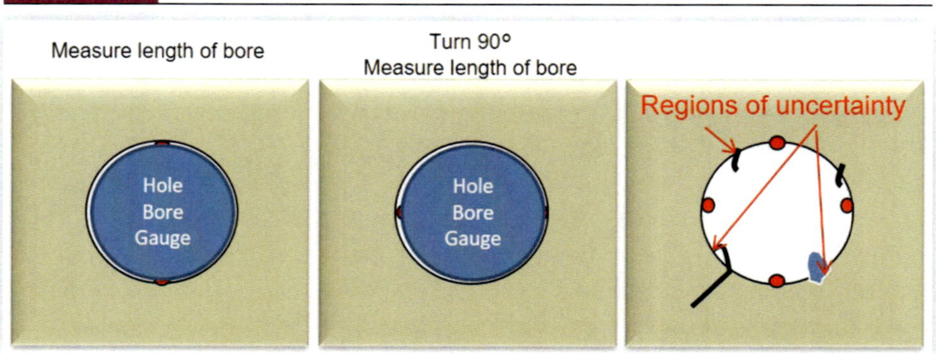

© SAE International/George Bullen

at 90° to the previous line of measurements. As shown in Figure 1.4, the assumption is made based on a controlled manufacturing process and the attributes of metal, in that the diameter of the hole-bore is acceptable (or unacceptable) based on measurements taken along the four lines of contact.

The "machine friendly" attributes of metal further translate to enhanced manufacturing efficiencies. Joining two pieces of the same metal together by welding or fastener is simplified by their common properties such as hardness, coefficient of thermal expansion, and chemical interaction. In the fastener example, drilling a hole through two pieces of the same or similar metal reduces the chance of hole-bore taper derived from the chip of a harder material being pulled through and out of the hole during the drilling process. Hole-bore diameters in stacks of aluminum over steel or aluminum over titanium are more difficult to control because as the harder steel or titanium material is encountered by the drill, the chip is pulled through the softer aluminum and cuts the material, enlarging the hole. In aerospace, the solution has been to step drill holes in both aluminum–aluminum, aluminum–steel, and aluminum–titanium stacks, thereby reducing the thickness of the chip and its ability to "chew" away the diameter of the softer material. Most applications are aluminum over aluminum in aerospace, but with the great migration of aerospace aluminum onto road vehicles with steel chassis, the chip issue will migrate to challenge automotive manufacturing. This issue/challenge is relatively minor in all past and current mostly aluminum airplanes since the aluminum–steel and aluminum–titanium stacks occur in a small percentage of an airplane. **Figure 1.5** shows the typical process for step drilling a fastener hole and then countersinking as a separate operation. The countersink is performed using a separate process and set of tools. In the manual process, a countersink with a pilot follows the previously hand-drilled hole and uses a gauged stop to control the depth of cut.

As automation technology has migrated into and become integral to the assembly processes for aerospace, the procedure of drilling a finished hole-bore with countersink "one shot" has become the norm. The rigidity of the machines, precision, and controlled

FIGURE 1.5 Step drill process with countersink.

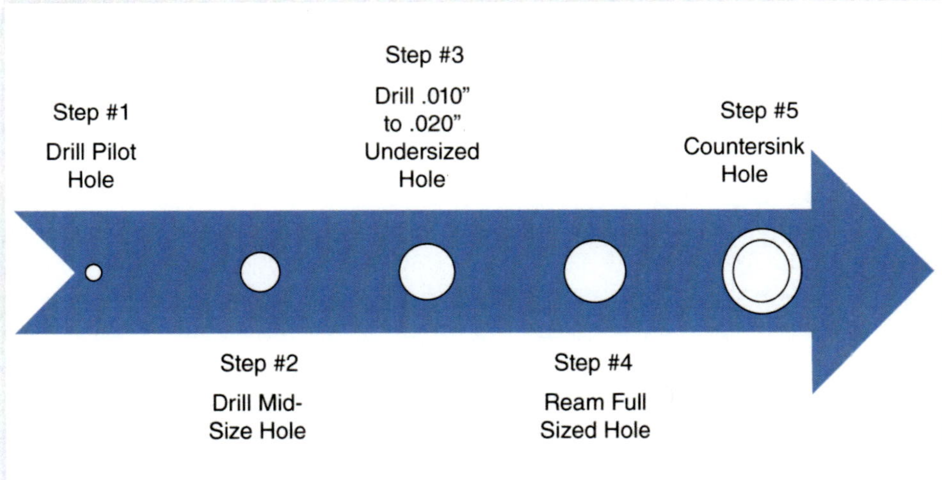

speeds and feeds has enabled the ability of manufacturers to drill precise holes with a countersink one shot through stacks of material. The drill/countersink algorithms that control the processes for these types of highly sophisticated machines are often proprietary and tightly held by the users and developers of the systems. Drilling and countersinking one shot through stacks of the same or similar types of metal material improves efficiency by eliminating step-drilling to a finished size hole-bore size and incorporates the separate countersinking operation into a single operation to finish a hole in preparation for fastener installation.

One of the advantages when drilling and countersinking one shot in metal-to-metal stacks is to leverage automation to install fasteners. The automated drill and fill option is available because of the tight fit, limited gauging, and uniform thickness of the material stacks. A tight fit ensures debris and burrs are not present between stacks, limited gauging is possible because of process consistency that enables spacing between hole-bore and countersinking inspections, and uniform and predictable thickness reduces the number of fasteners stored at the point of use. Limited fastener storage and consistent grip lengths increase the efficiency of the automation process through faster fastener selection, retrieval, and insertion. The process for drill and fill has yet to reach its full potential as the transfer of separate gauges to the automated process necessary to measure at the required intervals complicates and slows the automation. Many efforts have been made and ongoing development projects continue to try and transfer, adapt, and integrate previous hand-held gauges into the automated process. The array of gauges held in a tool changer needed to facilitate gauging, even at a 10% level, slows the drill and fill application of automation and often defeats the return on investment needed to implement an automated drill and fill solution.

1.2.1 Aluminum from Planes to Cars

High-strength "aerospace grade" aluminum is migrating into and becoming a larger contributor to the structural components of road vehicles. Road vehicle manufacturers are driven to reduce weight to increase fuel efficiency while retaining the safety and utility of a larger vehicle.

In the beginning of the automotive weight reduction efforts, size was the primary means automotive vehicle producers used to make lighter vehicles. Smaller vehicles use less material. Less material translates to a lighter vehicle. Consumers have pushed-back on automobile manufacturers and rejected small cars for many reasons such as reduced utility and safety.

It is a fact of physics: larger vehicles are safer than smaller vehicles. Smaller, lighter cars are generally not as safe as larger, heavier cars. **Figure 1.6** provides a consumer view toward the automotive manufacturers' drive to reduce size to meet fuel efficiency requirements through automobile miniaturization. The caption for this picture reads, "My Smart Car hit a squirrel. The squirrel refused medical attention and walked away unharmed."

Large vehicles have longer hoods and bigger crush zones, which gives them an advantage in frontal crashes. In studies conducted by the Insurance Institute for Highway Safety (IIHS), a heavier vehicle will typically push a lighter one backward

FIGURE 1.6 Results of a Smart Car accident. *"My Smart Car hit a squirrel. Thesquirrel refused medical attention and walked away unharmed."*

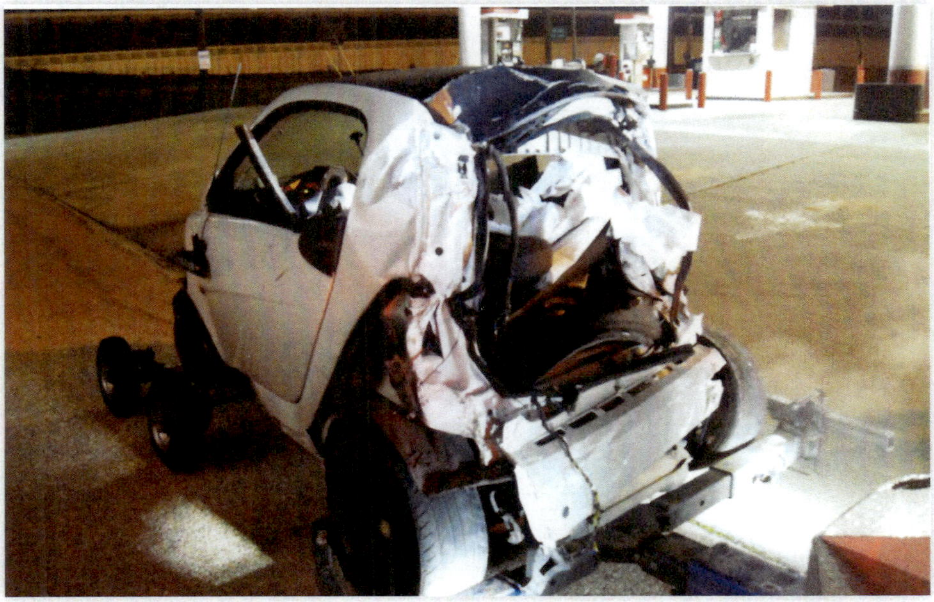

during the impact. Less force will be placed on the occupants of the heavier vehicle and more on those in the lighter vehicle, according to IIHS. Historically, the rates of driver deaths per million registered vehicles have been higher for the smaller and lighter vehicles.

Although aluminum reduces weight in larger vehicles, and the overall weight of the vehicle is reduced enough to increase fuel efficiency to meet the future federal guidelines, the vehicles would retain a weight advantage in a collision. The addition of aluminum or hybrid aluminum/iron materials would also lower the center of mass for these types of vehicles, because steel would remain a major component of the chassis. A lower center of mass relative to height would lower the center of gravity, make a more stable vehicle, and reduce instances of rollover.

Aluminum is one least-cost alternative to reduce weight across all vehicle types. While aluminum is a least-cost alternative material to reduce weight, it is not without transition-to-production challenges for high-rate production across all road vehicle types. Aluminum has been used in cars for many years in limited applications on body parts such as hoods, trunk lids, and tailgates. Some aluminum side panels exist on a few vehicle types. The application of aluminum to hoods, trunk lids, and tailgates represents attached parts that are bolted onto the main structure of the vehicle. The application of aluminum when combined with steel and integrated into the main structural components of a vehicle presents a significant design/engineering and manufacturing challenge to an industry deeply rooted in steel. To an industry characterized by factories filled with thousands of robots spitting sparks as they weld together steel pieces and parts, aluminum is an alien material fraught with unfamiliar manufacturing processes and challenges.

Aerospace has long recognized the advantages of aluminum to reduce weight while providing a more formable material than steel. The "goodness" of aluminum's characteristics is offset by its need for more complex manufacturing processes and its sensitivity to corrosion and damage during assembly. Aluminum can be welded, and in certain applications joined to other aluminum or steel parts and pieces; bonding and fastening are the preferred and most prevalent methods.

Complicating the engineering solution for transitioning a vehicle from steel to aluminum or from steel to a steel/aluminum combined assembly is the joining method. Joining steel-to-steel using automated welding technology is a well-understood manufacturing process with known outcomes within the automotive industry. Transitioning to bonding, fastening, and other means to hold all the vehicle pieces together presents a design and engineering challenge. The proper size, positioning, placement, and penetration depth of fasteners contribute to the strength, durability, and performance of a vehicle. Trying to simply transfer spot weld positions from a previous process to fastener placement is unacceptable. Also, complicating the design and engineering challenge of bonding and fastening aluminum pieces and parts together is the new manufacturing challenge for access and automation of the new processes required for aluminum integration into the vehicle. Automating the fastener installation process is much more complex than automating spot welding.

Fastening aluminum structure to other aluminum pieces or to steel adds complexity to the design engineering and manufacturing challenge. Corrosion is the enemy of aluminum. Aerospace has long known that given the smallest access to entry, corrosion in aluminum will find an opening and propagate. Corrosion in aluminum parts migrates, often in a clandestine way. Often it remains unseen and continues to degrade the strength of the material until part or assembly failure occurs. A fastened-together mostly aluminum vehicle has a thousand places for entry and propagation of corrosion, which can spread rapidly. Each fastener in aerospace, and by extension road vehicles, must be meticulously protected to prevent entry of corrosion.

Lowering the cost of integrating aluminum material into a previously mostly steel vehicle using mature and robust manufacturing processes will occur. As the application spreads and the learning curve matures, a robust and understood manufacturing process will evolve to incorporate the technologies, materials, and processes transferred from the aerospace industry.

The absence of a gauging solution to fully capitalize on automated drill and fill that has limited the full potential for automating this process in aerospace is magnified in the automotive production process. Airplanes are produced at a rate of maybe one-a-day. Cars are produced at a rate of one every 7 to 14 seconds. Slowing the automated process necessary for gauging every tenth hole is unacceptable with the current type of gauges. What is needed is a noncontact gauge that can measure the full depth and breadth of hole-bore sizes, countersink depths, and grip lengths in a single gauge that can be rapidly applied and integrated into the automated drill-and-fill process. If the gauge could also evaluate the joint surface between the mating pieces to ascertain and calculate the amount of moisture present, then corrosion evaluation could be rapidly integrated into the manufacturing conformance evaluation process for better vehicle quality.

1.3 Composites—Black Metal

Composites and specifically high-strength carbon fiber (HSCF) composites used in aerospace and rapidly migrating onto road vehicles have changed the manufacturing processes and gauging approach formerly used with all metal components.

HSCF are anisotropic materials whose properties are directionally dependent. Unlike isotropic materials such as metals that have material properties identical in all directions, the properties of anisotropic materials such as Young's modulus[1] change with direction along the object.

HSCF are fiber-reinforced plastic materials that consist of at least two materials: plastic that binds the fibers together, also called the matrix, and fibers, typically small in diameter and long in length. In general, the matrix imparts toughness, or crack resistance, and the fiber imparts ultimate strength. Composite material is much stronger in the direction of the fiber. If the fibers are unidirectional (all in the same direction), the composite material is strong in the direction of the fibers, but weak in the directions perpendicular to the fibers. This condition is alleviated by adding multiple plies laid with different fiber directions. The varying fiber direction in plies builds a laminate structure with strength in more than one direction. **Figure 1.7** shows the stack of materials common to building an HSCF part.

Each course or ply layer generates a "quilting" pattern as shown in **Figure 1.8**.

FIGURE 1.7 Varying fiber direction in plies during the layup process.

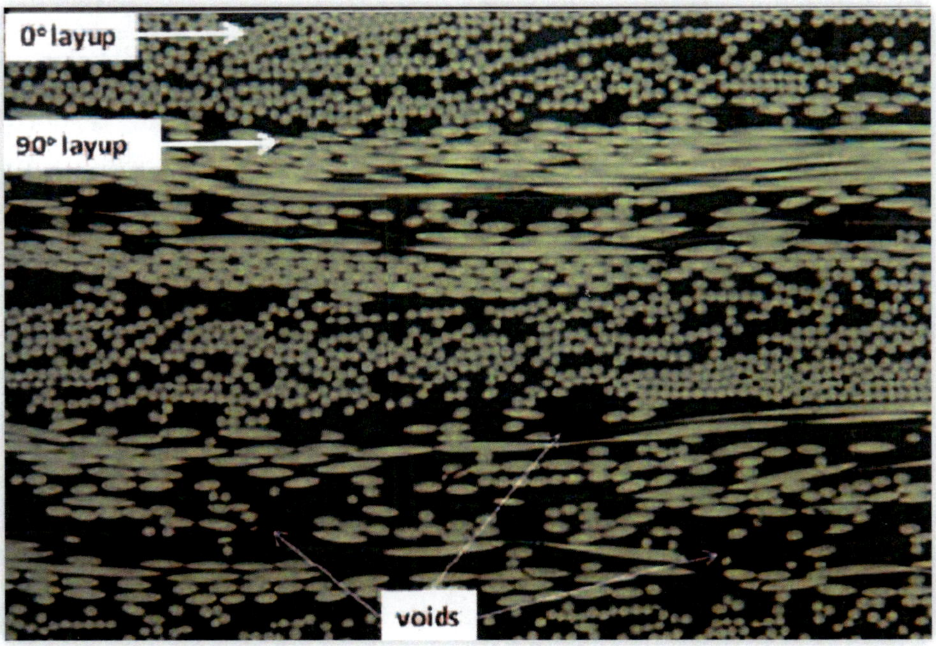

[1] Young's modulus is a measure of elasticity, equal to the ratio of the stress acting on a substance to the strain produced.

FIGURE 1.8 Quilting pattern for each course or ply layer.

© SAE International/George Bullen

The controlled manufacturing process for HSCF includes hand layup, compression of the laminate using vacuum, and resin permeation around fibers during consolidation and cure. The process creates parts that functionally meet the specified design properties. This same manufacturing process creates minute variations throughout the part. Each part (if produced correctly) is functionally the same but varies materially within the bowels of the parts composition. **Figure 1.9** shows an optical cross section of a portion of a part that demonstrates the variation and complexity of the internal composition of a cured composite part.

This internal variation translates to a highly complex and variable material when drilled and countersunk for joining pieces of an assembly together. Each time a drill bit or countersink penetrates and removes material from a composite part, the cutting edge is constantly moving through different strata that vary in composition, thickness, and fiber angle from hole to hole. Since the strength of a composite part is derived from its fibers and the ability of the plastic matrix to hold everything together, slicing and cutting through a composite material to create holes for fasteners becomes much more complex than the same operation through a metal material. Arguably, in many instances the process can cause crenulated fibers that diminish the strength properties of the part.

A crenulated fiber is one which has a wavy or scalloped outer radius, with the amplitude and frequency of the waviness resulting from the manufacturing process. The stress levels present with the crenulated fiber are significantly higher than for the circular fiber. In particular, large stress concentrations arise in the matrix tangential stress component at high crenulation amplitudes and high crenulation frequencies. The increased use and experience with HSCF composite materials has provided new insight into the complex interactions between fiber and matrix, specifically interactions that reflect on the reality of dis-bond [1-1].

FIGURE 1.9 Optical cross section of a composite part after cure.

Experienced manufacturers who fabricate and assemble composite parts have learned that the acceptance criteria and the gauges for metal are not transferrable from metal to composites. 100% of hole-bores and countersinks must be inspected. And 100% of the hole-bore and countersink must be evaluated to determine what effect the cutting operation has on the material. Simply gauging depth and measuring along four contact lines within a hole-bore is not sufficient. The current gauges meant for metal are supplemented by “eyes-on” evaluation after gauging. The “eyes-on” human action slows the assembly process. It also complicates and has defeated the ability to apply the complete drill-gauge-fill automated solution to composite assemblies.

Further complicating the assembly process is the condition of composite parts after fabrication. The layup process produces composite parts that have a tooled side and an un-tooled side resulting in a rough surface on the un-tooled side. Since the tooled side almost always faces outward for airflow and aesthetic reasons, the rough and uneven un-tooled or bag-side surface is mated against a precision machined metal substructure. This uneven mating surface must be filled with a liquid shim material to ensure a tight and pristine fit-up between the composite and metal parts. This complicates the drilling, countersinking, and gauging process. A drill bit, its associated speeds, and coolant must be multifunctional and designed to successfully penetrate a variety of dissimilar materials. Gauges must now measure and evaluate the impact of penetrating and removing material from dissimilar material stacks.

The impact of drilling composites alone is complex. When drilling composite to metal stacks of material, the process is complicated beyond the drill bits’ effect as it removes composite material. As it moves through and into the metal material, the hard chip generated must exit along the flutes of the drill bit and up and out through the composite material. The sharp edge of the chip becomes a cutting medium and can damage or enlarge the softer composite material.

Moving through a softer composite material, into a brittle liquid shim material, and then finishing a hole-bore in metal material can create separation and burrs between the mating plane surfaces of these three materials. A post-process must be performed to investigate, identify, and mitigate any drill-process impacts on quality. This post-drill process is called de-stack. The de-stack process involves taking all the parts of the assembly apart after drilling and evaluating the mating surfaces to assess, discover, and remove debris and burrs created that resulted from the drilling process. This non-value added activity is a direct result of the inability of current metal inspection gauge technology to fulfill all the necessary requirements to evaluate composite materials without human intervention.

Composite parts also vary in thickness. The thickness of the same composite part varies from part to part even when the manufacturing processes are tightly controlled. Therefore, each assembly is challenged by a family of parts that is not consistent each time they are positioned and joined together. This multivariate random variable is a primary cost and quality driver at assembly. It impacts form, fit, and function.

There are activities built into and necessary to the drilling, countersinking, and fastener installation process that are hidden within touch labor cost. One example is grip length measurement for fastener selection. Grip length incorporates safety factors for embedment and thread strip that are important both for the initial installation of the fastener and for long-term reliability for both loosening and fatigue resistance. Grip length assessment for fastener installation was always important but has become more important as composite materials have increasingly made their way into air and road vehicles. Composite parts complicate grip length determination, because their thickness varies by ±5% and in some parts by as much as ±8%.

In metallic assemblies, the mechanic confirms the grip length at every 10 or 20 holes, because the assembled thickness of metallic parts is very consistent. In assemblies that are composed of stacks of material incorporating composites, almost every hole is checked to determine the proper grip length before the proper fastener is selected. The process of sticking a grip length gauge into each hole to determine the proper fastener grip length is built into either the drilling standards or the fastener installation standards. It is a hidden cost and is time and quality dependent based on the discipline of the mechanic and their visual acuity when reading the gauge. This is an unseen cost that can be reduced or eliminated if grip length is calculated at the time the hole is measured, a capability current gauges do not have.

In the manual grip length assessment, every hole is measured as an independent event based on the composite variability assumption. Data are now not collected to compile patterns in each part geometry to determine areas of repeatable thickness. The mechanic measures the hole, reaches into the appropriate bin in the fastener kit, and installs the fastener. Advanced analytical data collection systems can collect discrete data and populate a knowledge base to perform pattern analysis and map each part for repeatability. This attribute would reduce the variety and the numbers of fasteners stocked at the assembly line and hasten the selection and installation of fasteners.

1.4 Summary

Specified hole/countersink conformance is generally a combination of position and D4&P. D4&P represents evaluation of dimension, depth, damage, debris, and perpendicularity. In aerospace vernacular, perpendicularity is referred to as "normality" or "normal to surface." The number of tools to assess and validate hole/countersink conformance is complex because of the breadth, depth, size, rate requirement, and maturity of the various assembly lines globally producing air and road vehicles and their components. The definition of the breadth and depth of hole/countersink assessment instruments is complicated by specific instruments used to evaluate each element of the hole. A countersink depth and angle conformance evaluation may use one instrument and a dimensional hole-bore evaluation another instrument, while yet another instrument is used to evaluate the overall depth of the hole-bore (grip length). The air and road vehicle manufacturing base and the finished air and road vehicle maintenance, repair, and overhaul industry is awash in the variety and complexity of the hole/countersink instruments used today. Some instrument manufacturers have attempted to combine many of the D4&P criteria into a single assessment instrument, but with limited success. The instruments generally have been expensive, complicated to use, and do not incorporate all the D4&P criteria into a single instrument. As more and more aerospace grade aluminum and composite materials are incorporated into road vehicles, the issues with fasteners, fit, and gauging long challenging aerospace manufacturers will migrate to road vehicles.

The devices now used have limited or no ability to collect data to populate a knowledge base for analytical assessment and evaluation to improve fit-up at assembly. When applied to automated systems, hole/countersink diagnostic instruments used nowadays offer limited dimensional evaluation and do not provide simultaneous one-shot drill/countersink and hole/countersink evaluation, or hole/countersink predictive analytics.

The introduction of composite materials into air and road vehicles has complicated the traditional hole/countersink assessment criteria because of its finished-part thickness variability, softer and dissimilar properties than the metallic substructure where it is mounted and attached, and the increased attention to other acceptance criteria such as fiber tear in the hole. The addition of composite materials further complicates the assembly process by adding a boundary layer of liquid shim between the composite piece (usually a skin) and the substructure. The liquid shim material is extremely brittle when cured adding a drill routine that must accommodate drilling through dissimilar stacks of material, concluding with "pulling" a metallic chip through a softer composite hole. Manufacturers are left with the inefficient manufacturing process of "de-stack" because mechanics cannot assess the interior condition of the composite hole such as fiber tear, damage to the liquid shim, and debris or burrs between the multiple stacks of dissimilar material. In the "de-stack" process, the manufacturer must remove the composite skin after drilling to evaluate the effects of drilling through the various stacks of dissimilar material. This step assures that the parts fit-up without interference from burrs and debris or that the drilling operation has not chipped, cracked, or otherwise damaged the liquid shim.

The requirement and opportunity for a device that can satisfy the full range of D4&P criteria and provide a full evaluation of the composite material and liquid shim

FIGURE 1.10 Torn fiber in a hole-bore.

© United Sciences

condition simultaneously, while collecting data to populate a knowledge base to facilitate advanced analytics would provide a new disruptive technology to air and road vehicle manufacturers.

As more parts made from composite materials have made their way into air and road vehicles, the design, definition, and macro manufacturing process effects of composite materials have become better understood but remain challenging. As composite material types have integrated into parts and assemblies for air and road vehicles, several micro effects within the manufacturing process have begun to manifest themselves and be recognized as contributors to decreased performance. Damage can occur at several scales within the composite material and structural configuration during the drilling operation. This ranges from damage in the matrix and fiber to broken elements, resulting in failure of bolted attachments. The extent of damage impacts load life and residual strength, therefore, it is critical to damage tolerance. Fiber breakage can be critical because structures are typically designed to be fiber dominant (i.e., fibers carry most of the loads). Figure 1.10 shows torn fiber in a hole.

The assessment of in-hole quality has become a critical component during the composite manufacturing process to ensure vehicle performance and longevity.

The chapters in ***Section One: Noncontact Ring Laser Metrology Systems*** describe in detail the emerging technology that replaces all previous gauges developed for metal applications now used for assessing hole-bore, countersink, and grip length. The described gauge replaces all the current gauges and combines the capability of multiple gauges into a single noncontact instrument. It provides the needed fidelity and resolution to evaluate the required elements resident in carbon fiber composite materials as well as metal. This single gauge solution closes the gap between automated drill/countersink and fastener installation by rapidly providing all the necessary evaluation criteria to validate in-hole conformance across the material stack spectrum without human intervention. It provides the solution to rapid drill and fill, formerly lacking in gauging devices and systems.

Reference

1-1. Carapella, E. E., Hyer, M. W., Griffen, Jr., O. H., Maahs, G., and Trice, F. "Micromechanics of Crenulated Fibers," Journal of Composite Materials. 28(14):1322–1346, August 1994.

CHAPTER 2

System Components

The solution often turns out more beautiful than the puzzle.

—Richard Dawkins

2.1 Introduction

Every new technology application has roots that grow out from a dissimilar technology when vision leads to innovation and creates extensibility into a new domain. The ring laser, noncontact gauge technology presented and described in this chapter is an example of technology first developed for soft tissue applications and meant to be used in the pristine environment of a medical office. The technology has been successfully modified and adapted for a completely alien and unintended use.

A company named United Sciences was awarded a Small Business Innovative Research grant from the United States Navy for ear scanner development to provide improved custom hearing protection needed to replace the one-size-fits-all foam hearing protectors that were being used. The improved hearing protection was needed to reduce hearing loss of flight deck crews and warfighters. Noise-induced hearing loss costs the United States military over one billion dollars a year.

The challenge to develop an ear scanner was substantial since the ear is one of the most difficult orifices to precisely measure. Every ear is different and measurement difficulties are compounded by the presence of wax, hair, liquid, and the absence of an available light source to illuminate the area to be measured. **Figure 2.1** shows the complexity of the human ear.

FIGURE 2.1 The complexity of the human ear.

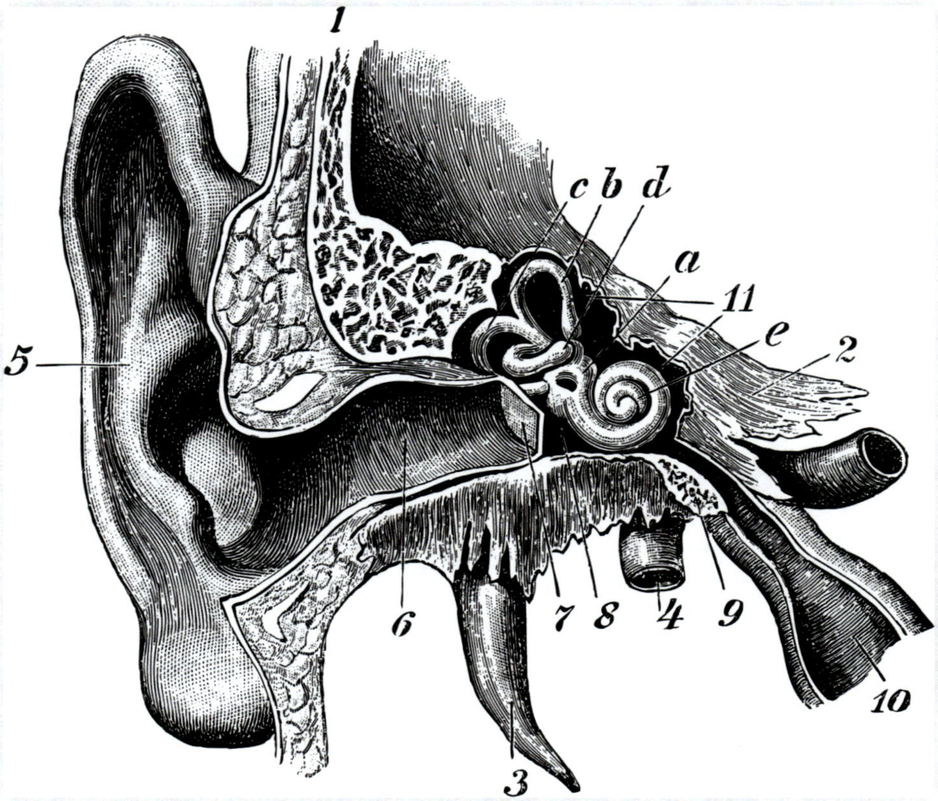

© Shutterstock

An in-ear "eFit" scanner was designed and developed to produce 3-D ear scans to create a 3-D printed product for optimal custom-fit ear noise reduction devices. The scanner won the Best New Product Award at the 2012 American Academy of Audiology's annual conference, AudiologyNOW. **Figure 2.2** shows the ear scanner as it is applied into the ear to capture 3-D data to be transferred to a 3-D printer for manufacture of a custom precision fit hearing protector.

FIGURE 2.2 The ear scanner as it is applied into the ear.

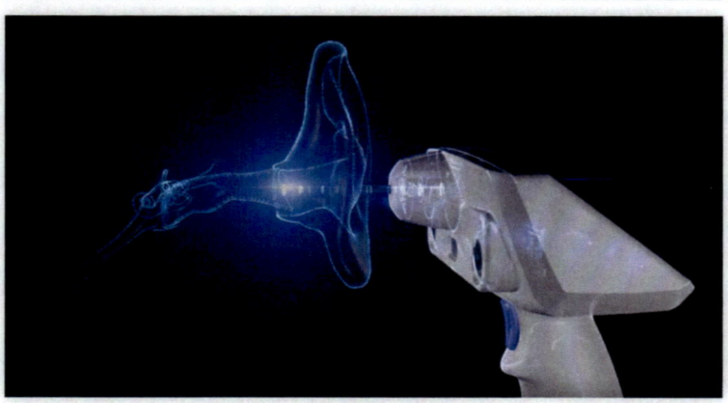

© United Sciences

Obvious and logical extensions from hearing protection to other applications led to the creation of hearing aids and wearables such as ear phones. The 3-D scans also enabled the consistent capture of unique biometric data in the ear such as electroencephalogram (EEG detection of brain waves), electromyography (EMG for muscle energy detection), and electrooculography (EOG for eye movement and tracking). **Figure 2.3** shows the basic scanning probe architecture and **Figure 2.4** shows how a custom-made, precisely produced ear device is fit into the ear.

FIGURE 2.3 The basic scanning probe architecture.

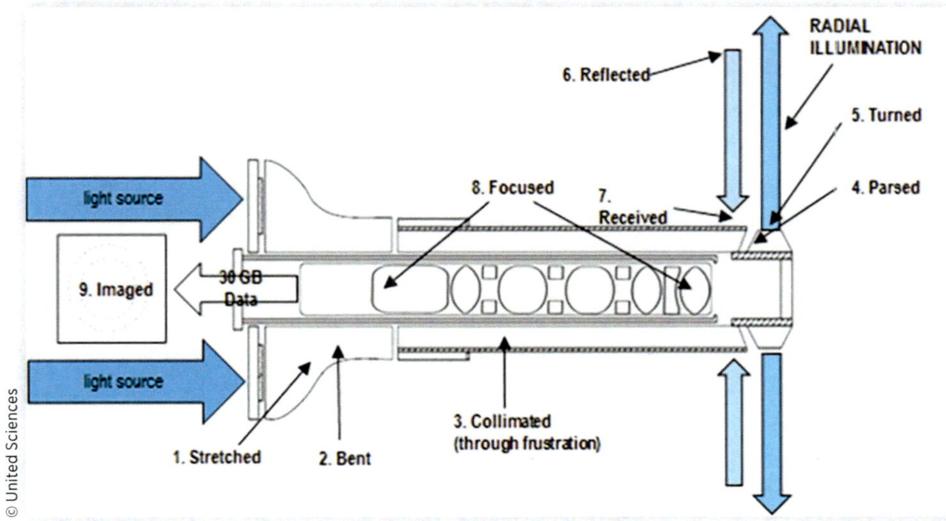

FIGURE 2.4 A custom-made, precisely produced ear device fits into the ear.

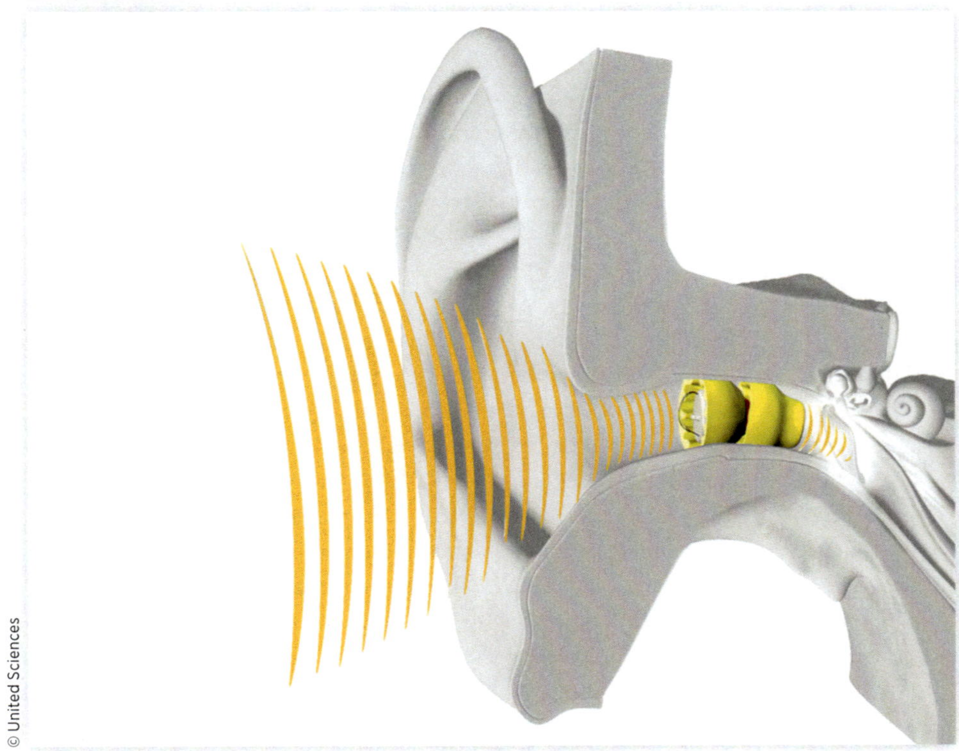

The roots for extensibility into aerospace, automotive, and rail vehicle applications are embedded in the lightweight precision noncontact medical measurement device for the internal and external features of the ear, and the data derived from the scan that are used to create precise custom in-ear devices, such as custom-fit hearing aids.

A Precision Engineering Division was launched by United Sciences in 2014 to focus on metrology and inspection of holes and countersinks in aerospace and ground vehicle manufacturing and assembly. The manufacturing-focused hole-bore diagnostic gauge development was based on the core components of the ear scanner. The core components are embodied in a noncontact semi-automated and automated hole-bore diagnostic gauge for sensing and evaluating hole-bore dimensional conformance for hole-bores previously drilled using hand methods, power feed drill motors, and automated drill/countersinking machines commonly used in the manufacturing industry. The gauge was designed to be manually or automatically inserted into a previously drilled and countersunk hole-bore.

The extension and adaptation of an ear scanner for transfer and use in the harsh environments of manufacturing for application on metals and especially high strength carbon fiber composite (HSCFC) materials was evolutionary and difficult. The perceived benefits, especially for application on holes drilled into HSCFC materials incentivized and drove the innovative designs necessary to adapt, modify, and overcome the weaknesses of a device originated for a more genteel environment. Overcoming the challenges of transferring a device intended for slowly measuring two ears in the quiet environment of a medical office into a factory environment was daunting. Consistently measuring hole-bores and countersinks in a noisy and comparatively dirty factory environment under the time constraints of a production rate where thousands and, in many cases, millions of critical dimensions need to be precisely measured in a timely manner was challenging. Normally, these challenges and others would present a chasm so deep and wide that investors and technology innovators would shy away and pursue technology transfers to areas that offered closer alignment with the core technology with less investment in resources and time. Challenges and frustrations began to emerge from within the aerospace manufacturing environment that incentivized investors and innovators to take the risk. These challenges and frustrations clarified the need for development of the core technology into a form factor that would be applicable within the aerospace manufacturing process.

Timing incentivized the adaptation and invigorated the vision and funding necessary to realign the purpose of an ear scanner into an airplane scanner. The aerospace industry was ripe and ready for a single gauge that could perform the multiple tasks that required and still require several gauges. The use of a single gauge for various tasks, and the use of a single gauge for a variety of hole-bore sizes would significantly reduce the complexity, cost, and effort of measuring millions of holes, and reduce sustainment and maintenance of a large array of gauges. Further incentivizing the adaptation of the ear scanner technology for measuring manufactured products was the increasing realization of deficiencies inherent in gauges developed for metal but now increasingly used for HSCFC. As the design for adapting the ear to airplane scanner matured, it became clear to the sponsors and stakeholders that the emerging benefits were far beyond the initial perceptions for enhanced quality and efficiency. The attributes of the ear-to-airplane gauge began to be viewed as a panacea for mitigating many of the difficulties of airplane assembly and airplane assembly automation. And the ear gauge modification for manufacturing applications was also positively viewed by auto and rail vehicle producers for the same reason the aerospace industry became increasingly interested.

The key system elements described in this chapter provide the core components that empower this technology to perform all the tasks necessary for transfer from ear to airplane, road, and rail vehicle manufacturing applications. Airplanes and to a lesser extent automobiles and rail type vehicles are highly complex machines that must operate reliably in harsh environments. The performance of the interdependent array of technologies harmonized for their successful operation is critical to their mission. Airplanes, automobiles (including other road vehicles), and rail vehicles reliably deliver humans and the goods that sustain their existence reliably and safely to their destinations throughout the world. Road, rail, and air vehicles are critical to a web of interconnected and complex distribution systems that sustains human existence in our modern world.

Airplanes, road, and rail vehicles are dependent on equally complex manufacturing specifications, controls, and technologies that ensure the vehicles critical to our continued survival are fabricated and assembled to meet their intended performance requirements. They are manufactured using a variety of processes that require unique gauging solutions that include access, precision, frequency, repeatability, and variety. A highly complex global manufacturing environment, especially for road, rail, and air vehicles has evolved that relies on a common core of linked interdependencies that culminate at final assembly points for each product type. Without a common set of clear and concise specifications resulting in an agreeable and verifiable method of validation, pieces of the final product would arrive at assembly points without the ability to be dimensionally fit together.

What makes the technology described in this section of the book so attractive to the global road, rail, and air transportation manufacturing and sustainment stakeholders is its core technology capability. The resident core technology capability is the genesis for a single gauging solution to replace multiple gauges (simplification) while enabling linkage throughout the global digital tapestry for communication along the manufacturing value stream. As-manufactured dimensional connectivity elevates and links the globally disparate manufacture of pieces, components, and subassemblies into a unified global enterprise for reduction of variability. Variability reduction eases fit-up at assembly, reduces cost, elevates quality, improves efficiency, and provides a higher degree of reliability. Better reliability lowers sustainment cost.

Sustainment cost is lowered by the ability of the ear scanner's core technology to capture an as-manufactured digital twin of the manufactured part. The digital twin is used to carry as-manufactured data with the part until its end of life. In the later chapters of this section, the digital twin attributes and contribution to reducing sustainment cost will be addressed and described.

The core technologies and components and their basic contribution to the single gauge solution are described in this section. The descriptions of the core component are necessary to understand how the entire system operates.

Core components are presented that are essential to the operations of the gauge system. The core components are highly adaptable. Examples of specific applications are presented to illustrate how various users have applied the core technology within the confines of their specific needs.

2.2 Optical Probe

Most breakout and disruptive technologies have a core component or set of components that need to reside in the new application to realize the benefits derived from their use. This may seem obvious but modification or elimination of key elements of the core components can result in performance degradation or failure of the new application. Other hardware, software, housing(s), technology, and electronics can be added or reconfigured, but the core component must remain intact to realize the value resident in the disruptive technology. The ring laser scanning technology adapted from the ear scanner for application into aerospace manufacturing called Aeroscan™ requires the key ear scanner core technology to be viable. The probe assembly composed of the optical probe, probe-to-camera mount/attach ring, and camera are core technologies that must be precisely fabricated, assembled, and calibrated for the technology to function properly in the newly targeted domain. Without the core technology and its precise coordination and calibration during the fabrication and assembly process, the accuracy and functionality of the entire gauge is defeated for aerospace, automotive, and rail precision applications.

The optical probe assembly is shown in Figure 2.5. The probe housing is composed of an optically pure glass with an outside diameter of 3.2 mm or 126 in. The glass housing is flared at the base for attachment to the camera portion of the probe assembly. A threaded attachment ring compresses against the flair and holds the glass housing tight against the camera portion of the probe assembly.

The laser-based noncontact measurement probe utilizes radial illumination and is a non-spinning probe. The probe does not spin and it has no moving parts; the entire gauge is moved along an axis by a linear slider with its position monitored through a closed-loop feedback system provided by precision scales. The probe does not have to be perfectly centered (can be decentered by 0.250 mm without a significant change in measurement). The fact that the probe assembly operates without spinning provides

FIGURE 2.5 Basic Aeroscan™ probe configuration.

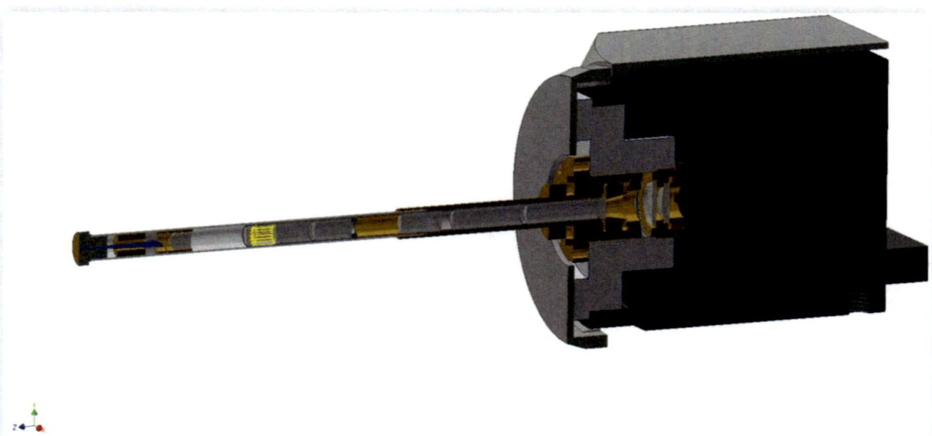

© United Sciences

an attribute that increases measurement and processing speed (time in the hole) and reduces the strain on wiring that would have to be wrapped and unwrapped as the probe entered and exited the hole-bore.

The gauge uses a self-actuating 3.2-mm (0.126 in.) outside diameter (OD) optical probe and laser delivery system to illuminate and image the interior of a drilled hole-bore for accurate and repeatable hole-bore measurement and diagnostics. The laser delivery system is engineered to project a ring of light within the hole-bore for imaging. The laser light is specifically tailored to radially illuminate and image both metal and composite materials, as well as metal-composite stacks of materials. Tested materials include titanium, aluminum, stainless steel, HiTuf™ steel, and most types of HSCFC material. The probe image and processing software can differentiate between different types of materials. For example, the intersection between a stack of carbon fiber composite material joined to a piece of aluminum material can be precisely identified with the image clearly showing the carbon fiber and the aluminum as separately identifiable images. The system is able to measure hole-bore diameters and countersinks in a mixed material stack of any order of arrangement, including up to five layers of mixed materials of aluminum, carbon fiber-reinforced plastic, and titanium. The imaging can also show and measure any gaps that might exist between the pieces of joined material. The interface gaps of 0.002 in. or greater can be measured at the interface of the mating parts making up the material stack being measured, to a repeatable accuracy of ±0.0005 in. The current Aeroscan™ system is capable of measuring through material stack thicknesses of 2.25 in. (57.15 mm).

The system is able to measure holes of a diameter as small as 0.1875 and 0.850 in. The system is able to measure countersinks in the range of .200–1.000 in.

The probe also includes a wide-angle low-distortion lens structure. The lens structure focuses the direct reflection of the radial light delivered into the hole-bore ring onto the camera portion of the probe assembly. The images captured on the camera are used to image the interior of the hole.

The probe is moved into the hole-bore without contacting either the hole-bore itself or debris within the hole-bore. The probe and the images created through it are not corrupted by contact with the hole or debris. This makes the probe especially useful in measurement and diagnostics holes drilled through composite materials. The probe and other components are sealed against contaminants and are cleanable with solvents common to the aerospace manufacturing processes. The probe capability has been certified for cleaning using a 2-h emersion test in various solvents commonly used in the aerospace and automotive industry. Emersion tests were performed using chemicals that included denatured alcohol, isopropyl alcohol, methyl propyl ketone, and Skykleen™.

The probe provides a universal solution for a variety of hole-bore sizes and depths. The probe can be placed into, measure, and provide diagnostics for holes ranging in size from 3/16 in. (4.8 mm) to ¾ in. (19 mm). A single one-gauge solution can be applied to most airframe holes. Figure 2.6 shows the radial illumination as it projects a ring of light into a hole-bore.

FIGURE 2.6 Radial illumination as it projects a ring of light into a hole-bore.

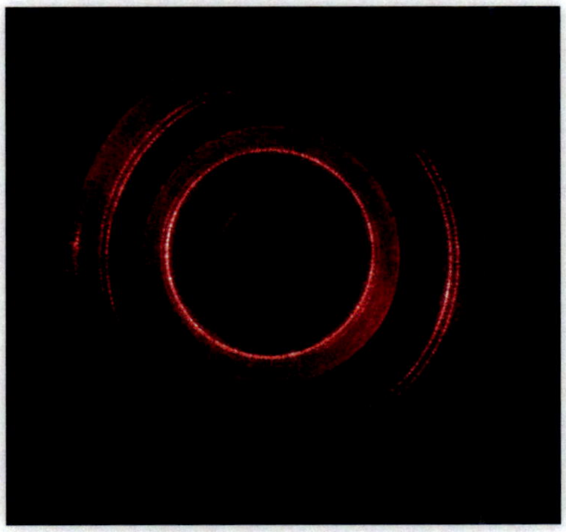

© United Sciences

FIGURE 2.7 Calibrated probe to attain hole-bore perpendicularity.

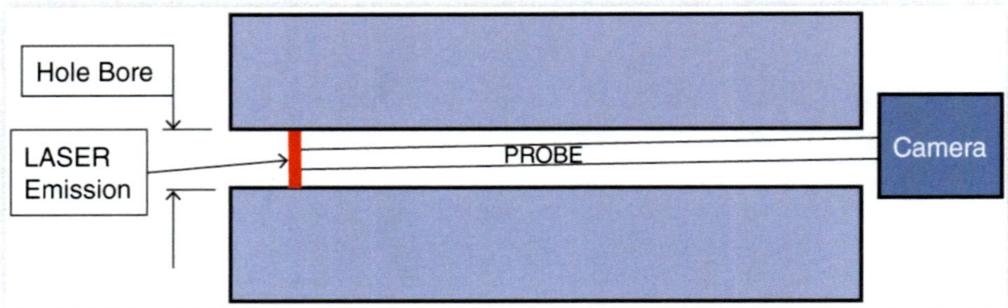

The challenge for precisely fabricating and calibrating the probe is driven by the variability of the inside diameter (ID) of the glass lens containment tube. The variability of the ID complicates the precise positioning of the lenses needed to project the ring of laser light perpendicular to the centerline of the probe. A ring "canted" off-center causes a reflection that geometrically reduces the accuracy of the gauge as the reflection ring angle-of-error increases. Calibration must align the laser ring projection to the centerline of the holes to be measured. **Figure 2.7** is an exaggerated illustration of a calibrated probe and its angular relationship to the hole-bore to achieve center of hole-bore perpendicularity.

Further complicating the calibration process is the variability of the flared end of the glass probe tube. Perfectly aligned lenses along the centerline of the glass probe tube will project a ring of light perpendicular to the tube until the probe tube is fastened to the camera by means of the camera mount ring. Since the bottom of the flared probe skirt is clamped tight and sealed to the camera lens, any deviation of the flared probe skirt from 90° to the centerline of the probe tube will cause angularity in the laser ring projected against the hole-bore. Both the variability of lens assembly and the variability of the flared probe skirt necessitate calibration either through mechanical means or software to align the axis of motion through the hole-bore to the ring laser projection so the gauge sustains its accuracy throughout the measurement cycle.

The anomalies combined with the glass composition of the probe have been identified as areas of focus for improving the Aeroscan™ gauge. Calibration of each Aeroscan™ probe assembly can take from hours to days, driving manufacturing costs. It also makes the gauge sensitive to factory operational environments where vibration can affect the calibration. However, vibration tests were recently performed at Boeing Airplane Company in Everett, Washington, DC, USA, without any degradation in accuracy or repeatability of the tested gauges. The Aeroscan™ gauge was robust enough to survive forces of 10 g at 10 Hz in the x, y, and z directions for 5 sec per hole during the fastener installation cycle. G load testing was not performed during the hole-bore measurement cycle.

The glass probe is also vulnerable to breakage. The vulnerability of the glass probe has been identified as both a benefit and a liability. The sensitivity of the glass probe to impact (breakage) is a concern. Mitigating the concern, especially in an automated environment, is the realization that a glass probe will shatter if it impacts the composite skin of an airplane causing little or no damage. Current gauges made from metal can and have caused damage when the same impact occurred. The robust gauge made from steel might survive the same impact but cause a defect to the product. The preference therefore is to sacrifice the gauge when impact occurs. Evaluating the impact damage to the airplane skin and performing a repair are more expensive than replacing a broken probe. The fragile nature of the probe drove a design solution for rapid probe replacement approaching 1 min. The 1-min replacement of a broken probe is possible using a mechanical type lock/unlock device that precisely positions the new probe without need for recalibration.

2.3 Camera

The Aeroscan™ gauge uses a laser/camera system to scan the inside of a drilled hole/countersink. Figure 2.8 shows the laser ring, optical glass, and lens stack (previously described) mounted onto the camera portion of a probe assembly by means of the threaded probe mount. The threaded probe mount is brass material. The camera and probe mount are protected from debris by an attached 3-D printed cover.

FIGURE 2.8 Laser ring, optical glass, and lens stack (previously described) mounted to the probe assembly camera.

The camera is necessary to capture the images that the laser reflects through the lenses. The ring of light is produced in the hole and reflected down through lenses to a camera. The camera images the ring at 32 frames per sec (FPS) and a computer reconstructs the image of the laser image into a cross section of points in XY-space. The image contains ~340 data points in each cross section. A full 360° point cloud is not possible (now) because the wires that power the light diode at the tip of the probe run along the outside of the probe tube obscuring the reflected image. The wires that power the diode are connected to a power source behind the camera. The wires obscure approximately 20 data points in the point cloud. One data point is taken at every degree of arc. While some criticism has been articulated regarding missing 20° of data, the data points provided by Aeroscan™ far exceed the 4° of data (one row at 90° of rotation along the hole-bore) that current gauges provide.

Diameter measurements are calculated for each cross section. As the probe moves through the hole, the system obtains cross sections at a rate determined by the desired fidelity of the user. The cross section is similar to slices taken by magnetic resonance imaging. The closer the cross sections are, the greater the fidelity of the "slices." For example, at 32 FPS, a one-inch thick hole-bore could be scanned in 6.25 sec if an image "slice" was desired at every 0.005 in. The speed to scan a hole is determined by how finite the user wants to evaluate a hole-bore and countersink. Figure 2.9 shows a hole-scan point cloud at a resolution of a "slice" at every 0.005 in., and shows the void created by the power wires.

Less than a second after the laser exits the hole, measurements are made on the entire hole-bore data set to produce countersink depth, bore min/max diameter, and grip length data, gap analysis, burr presence at the interface and backside, and hole angularity. Hole-bore angularity is used to calculate the possibility of a gapped fastener.

Several types of cameras have been developed to facilitate various housings that the probe-assembly must fit within. Figure 2.10 shows two cameras side by side where the

FIGURE 2.9 Point cloud from scan taken every 0.005 in.

FIGURE 2.10 Two cameras side by side where the top probe-assembly camera shown is much smaller than the standard camera.

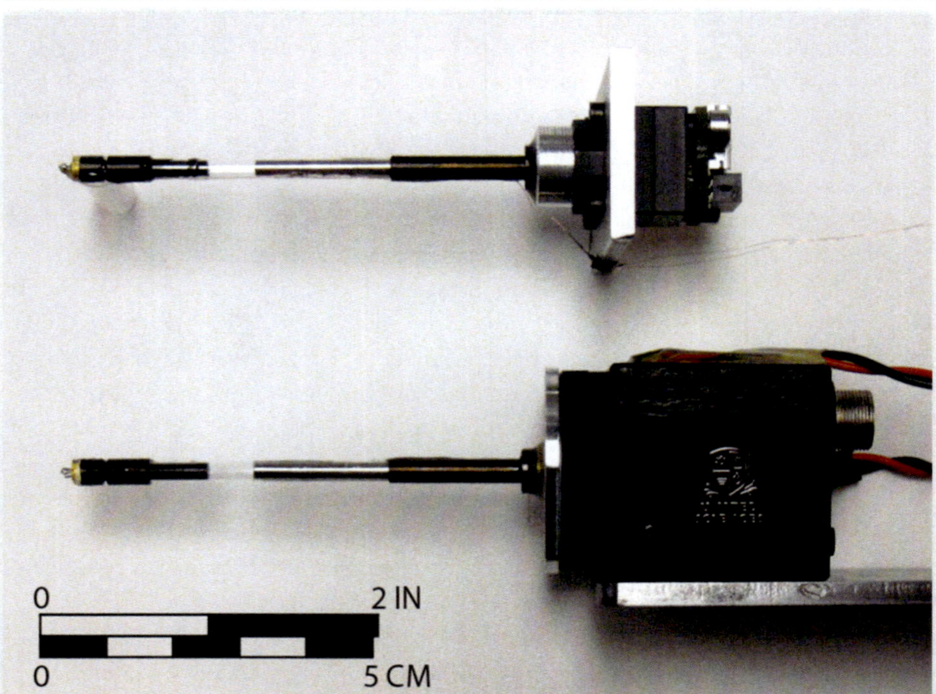

top probe-assembly camera shown is much smaller than the standard camera where the design team worked with a customer to redesign the form factor to fit within the end effector of the customer. Some sacrifice in speed was made as the smaller camera has a slower imaging speed (in FPS).

While the camera is smaller, the functionality remains the same to facilitate and enable the core technology and purpose of the probe assembly.

2.4 Probe-Camera Assembly

The probe-camera assembly is moved in and out of the hole-bore by means of a linear actuator. A linear actuator is an actuator that creates motion in a straight line. Mechanical linear actuators are used to guide the probe assembly. As previously mentioned, it is essential that the light emitted from the probe ring light source be perpendicular to the centerline of the hole-bore, and that the line of motion into and out of the hole-bore has to be parallel to the centerline. The motion can travel offset to the centerline of the hole-bore, but ideally run parallel to the centerline.

As a result of the variability from the lens and lens tube flare within the elements contained in the probe assembly, a probe-assembly mount is used to join the calibrated probe assembly and the linear actuator. The probe assembly to linear actuator mount contains attributes that enable the alignment of the calibrated probe assembly along the centerline of the linear actuator. The alignment provides datum plane coordinates for positioning the tip of the probe to intersect the centerline of the hole-bore to be measured and adjusts the normality to ensure the probe travels in and out of the hole-bore parallel to the hole-bore centerline. Figure 2.11 shows the hole-bore probe assembly with the hole-probe-assembly adjustment mount directly behind.

The development of a lens tube for the probe where the ID is tightly controlled will eliminate the variability when placing the lenses into the tube. Lens variability control will mitigate the lens angularity error and, in turn, eliminate one contributor to the angle error in the ring laser projection. Grinding the bottom of the lens tube flare perpendicular to the centerline of the lens tube will eliminate the second contributor to a canted laser ring light.

By performing these two manufacturing changes, the probe-assembly specific calibration requirement is eliminated, the probe assembly adjustment mount is simplified, and the probe can be rapidly changed or interchanged easily with other probes.

FIGURE 2.11 Arrow points to the hole-probe-assembly adjustment mount directly behind hole-bore probe assembly.

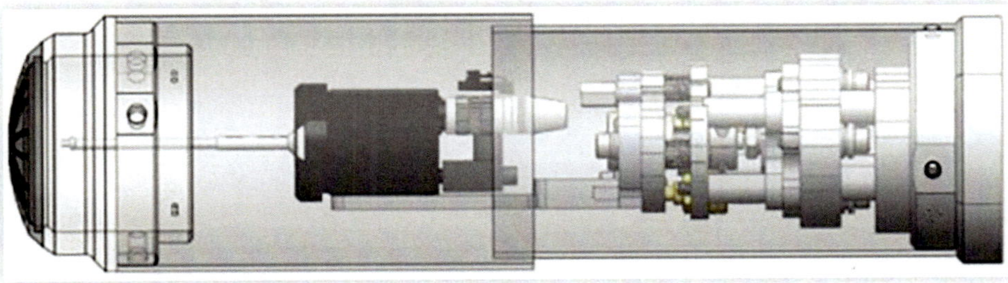

2.5 Machine Vision Software Package

The gauging system includes a machine vision software package for hole-bore imaging and diagnostics. The machine vision software provides real-time hole-bore diagnostic assessment, complete hole-bore imaging, predictive tool life-cycle analysis, and long-term knowledge-based analytics. The software identifies the primary reflection of the ring of illumination light with extreme precision, filters all secondary reflections, and uses the filtered image of the primary reflection to identify both hole-bore and countersink health and conformance assessment while creating a 3-D representation of the interior of the hole. The software can identify from the 3-D representation of the hole-bore predictive tool wear and analytics of the hole previously thought to be cost prohibitive. For example, the software is capable of identification of micro-delamination in composite materials, burrs, rifling, gaps, and many other anomalies of the interior of a hole that is now either ignored or identified in a cumbersome and costly manner.

The Aeroscan™ gauge is controlled by the Aeroscan™ 4.1.exe software application that runs on a Windows computer. The application needs to be started for the gauge to operate.

Aeroscan™ returns all requested attributes for drilled holes and communicates the attributes to a host controller within 1 sec of completion of the hole-bore measurement cycle using a PROFINT communication system. The product notional representation is contained in Figure 2.12.

FIGURE 2.12 Aeroscan™ product notional representation.

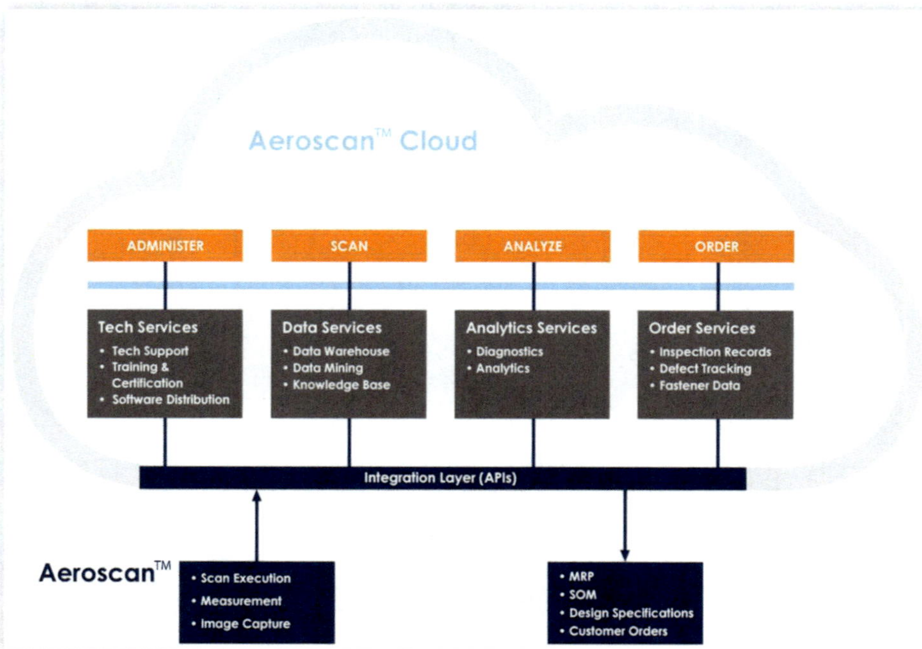

The following are needed to operate Aeroscan™:

1. One PC 1 Logitech keyboard dongle
2. One VGA cable (optional)
3. One LCD monitor (optional)
4. One keyboard (optional)
5. One power cable
6. One Ethernet cable

2.6 Hole-Bore Location Guidance System

The gauge has a hole-bore location guidance system that directs an operator to precisely align the gauge over the hole. The hole-bore location guidance system has an alignment camera that provides a live video stream of the alignment of the gauge and projects cross hairs upon the live video stream. An operator visually aligns the cross hairs upon the center of the hole that needs to be inspected. The live video stream is magnified such that an operator can easily align the gauge over any size of hole-bore, small, or large. Upon the precise alignment, an operator presses a single button to affect an entire scan of the hole. The gauging software automatically actuates the probe, images the hole, and provides 3-sec real-time measurement and diagnostics of the hole.

The housing consists of two pieces: an upper and lower piece designed to be interlocking and precisely joined together without need for assembly fixture(s) by means of eight attachment screws.

The upper housing is a closeout housing that holds the activation switch, and two LED lights for operator information. The activation switch and the LED lights are wired to the components contained in the lower housing. Two LED lights are located next to the activation switch and are used to notify the operator when the gauge is self-calibrating (red), when the gauge is ready to operate, and when the gauge is operating (green).

The lower housing contains all the components, parts, mechanisms, and wiring for the unit operation.

Also attached to the lower housing is one quick disconnect connector containing all pins to facilitate rapid removal and replacement of the gauge from the computer interface. The connector is wired to all components contained within the lower housing for power and communication for gauge operation.

The upper and lower housing are molded or 3-D printed to facilitate manufacture and positioning of the internal operating components, parts, wiring, and mechanisms for the systems operation and interface to the external computer system. A removable nosepiece is also attached to the lower housing.

The internal components located within the lower housing consist of

1. A probe assembly and mount
2. A probe assembly mount-to-motor interface
3. A saddle carriage that the probe assembly mount is affixed to for providing a means of precise conveyance of the probe assembly into and out of the housing
4. A stepper motor to precisely move the probe into and out of the housing through the nosepiece
5. An electronic measurement feedback scale to confirm positional accuracy during the measurement phase of the probes conveyance
6. All internal wiring

2.7 Summary

At the heart of every disruptive innovative technology are one or more core technologies that may be transferred into another application within a completely disparate technology that is also disruptive.[1]

The core technology of radial illumination led to a succession of inventions that grew out of the goal to simply make a better ear plug to improve hearing protection. The path for the progressive evolution of this technology was logical and proceeded with additional devices such as ear buds for phones and music type playing devices. The willingness by investors to take on the risk of moving from a completely different technology application into the harsh environments of the factory would not have occurred if not for timing. Aerospace was beginning to recognize that the material of the future (HSCFCs) could not be measured properly with the gauges that were transferred from the metal "world" of manufacturing. And the gauges meant for metal could not assess all the necessary requirements for validating the specified engineering requirements.

Compounding the deficiencies of gauges meant for metal was the quantity of gauges needed to measure each hole and each hole-bore size. What the aerospace, road, and rail vehicle manufacturing base needed was a single gauge to perform all the measurements required of gauges meant for metal applications and with the added capability to capture data necessary to assess and validate (or reject) HSCFCs, all contained in a single gauge.

The ear scanning hearing protection business spawned a technology that offered the potential to meet the demand driven by the aerospace industry for a new gauge that would meet their one-gauge requirement. The market draw, potential market size, and innovation drove the development of a disruptive technology to meet the demand. The challenge drove a 2-year journey and several million dollars of investment to transition a soft tissue gauge to be used "occasionally" in a pristine environment into a robust measurement gauge meant for a manufacturing factory environment. A gauge that would be relied upon to continuously measure thousands of holes a day in a factory

[1] A disruptive innovation is an innovation that creates a new market and value network and eventually disrupts an existing market and value network, displacing established market leading firms, products, and alliances. The term was defined and the phenomenon analyzed by Clayton M. Christensen beginning in 1995.

environment having an ambient temperature range from 60 to 110 °F (15 to 43 °C), and an ambient relative humidity range from 10% to 95%.

This chapter has described the core technology resident and critical to the operation and function of the new gauging system. The gauge has three primary components: the optics module (probe, probe mount, and camera), a magnetic linear encoder read head, and circuit boards.

The Aeroscan™ uses a laser/camera system to scan the inside of a drilled hole/countersink. A ring of light is produced in the hole; the camera images the ring, and the computer reconstructs the image of the laser into a cross section of points in the XY-space ~340 data points in a cross section. Diameter measurements are calculated for each cross section

As a slider drives the probe through the hole-bore while the system obtains cross sections, depth information comes from the linear encoder. A 3-D model of the hole is generated by combining the cross section points and depth data.

Less than a second after the laser exits the hole, measurements are made on the entire hole-bore data to produce countersink depth, bore min/max diameter, and grip length data.

CHAPTER 3

Operation

It doesn't stop being magic just because you know how it works.

—Terry Pratchett, The Wee Free Men

3.1 Introduction

The ear scanner ring laser noncontact measurement system was developed specifically for the aerospace industry and includes

- Optimized optics for aerospace hole-bore sizes
- A laser delivery system optimized to illuminate and image the materials used in the aerospace industry
- Aerospace hole-bore measurement, diagnostics, and analytics software
- A robust manufacturing and field calibration system
- A hole-bore location system that makes the aerospace developed adaptation readily implementable on the manufacturing floor.

3.2 Measurement Capabilities

The gauge is capable of accuracy for many measurements including hole-bore dimension, debris, damage, depth, and perpendicularity (D4+P). For example, hole-bore diameter is measured within 0.0002 in. Debris and damage are imaged in real time with accurate

FIGURE 3.1 Hole-bore gauge measurement terminology.

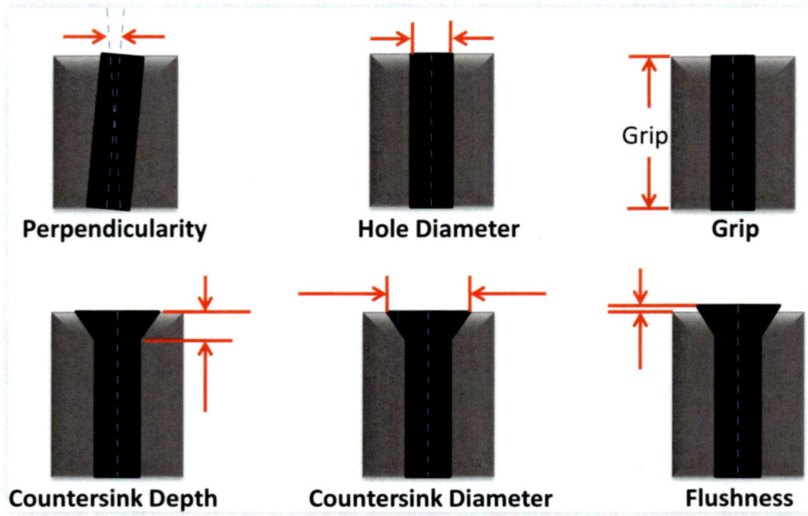

© SAE International/George Bullen

identification of the impact of drilling artifacts such as excess liquid shim, resin in the hole, and burrs, including exit burrs, gaps, cracks, and excess dust. Perpendicularity and grip length are measured with accuracies far exceeding industry standards. Depth or grip length is measured to accuracies within ±0.001 in. Composite-specific measurements are also identifiable with the gauge such as pre-de-stack analysis, micro-delamination, gaps, and fiber tear-out. Figure 3.1 shows hole-bore gauge measurement terminology.

FIGURE 3.2 Example of rifling in the hole because of wear in the drill bit.

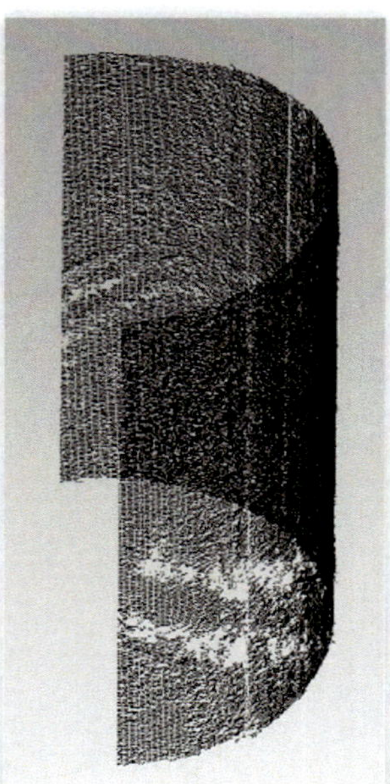

© United Sciences

3.3 Diagnostics Capabilities

One gauge provides real-time hole-bore diagnostics, including determining whether a hole is within specified tolerance for D4+P. The repeatability of the gauge eliminates human subjectivity in evaluating hole-bore diagnostics, enabling consistency across inspectors and inspection locations.

3.4 Analytics Capabilities

A database is maintained of all measurements, diagnostic results, and a complete image of the hole-bore. The data warehousing enables predictive analytics of tool life, hole-bore forensics, feedback through maintenance and repair operations, and long-term knowledge-based analytics for improved manufacturing tolerances and manufacturing processes. The Aeroscan™ gauge operating system is a closed-loop analytic solution that enables the actual impact of its use to be quantified. Figure 3.2 shows the rifling in the hole because of wear in the drill bit.

3.5 Composite Capabilities

The gauge is especially well suited for measurement, diagnostics, and analytics of holes in composite materials that add additional considerations to hole-bore inspection. Parts made from composite materials have thickness variability, are softer, and have dissimilar properties than the metallic substructure on which they are mounted and attached. Additional issues such as micro-delamination, the need for de-stack, and in-hole fiber tears also complicate the composite part assembly process. Composite materials have a boundary layer of liquid shim between the composite part and the substructure that can lead to excess debris between the parts or in the hole. The gauge is noncontact, so it is not affected by debris within the composite hole and optically images the hole allowing unique composite artifacts to be analyzed.

3.6 Speed and Repeatability Capabilities

Measurements and diagnostics are provided in real time. Measurement, diagnostics, and analytics are available for each hole within seconds. Repeatability of measurement-to-measurement is measurable to 0.00001 in. Repeatability is derived from a machine vision and imaging capability that does not require precision application by the operator. Repeatability is highly accurate providing repeatable measurements of the hole-bore as long as the probe does not touch the hole-bore wall.

3.7 Strategic Impact

The strategic impact of one gauge for both handheld and automated configurations is significant and far-reaching. The impact stretches beyond its complete hole/countersink diagnostics, composites evaluation, and analytics capability. It is a complete solution for the air vehicle original equipment manufacturer (OEM), their clients, and supply chain to provide one common inspection and measurement solution. One gauge provides a hole/countersink diagnostic instrument that can span the entire air vehicle supply chain to provide commonality, connectivity, communication, and integration. **Figure 3.3** shows the total impact of a one-gauge solution across the total product life cycle for OEM, client, and supply chain.

The functionality is the same whether applied by hand or by machine, providing a common instrument and process that is not present in other handheld and automated devices. Current hole/countersink instruments use different devices for handheld and automated hole/countersink assessment, which prohibits integrated data and information transfer across platforms and instruments for factory

FIGURE 3.3 Total impact of a one-gauge solution across the total product life cycle for OEM, client, and supply chain.

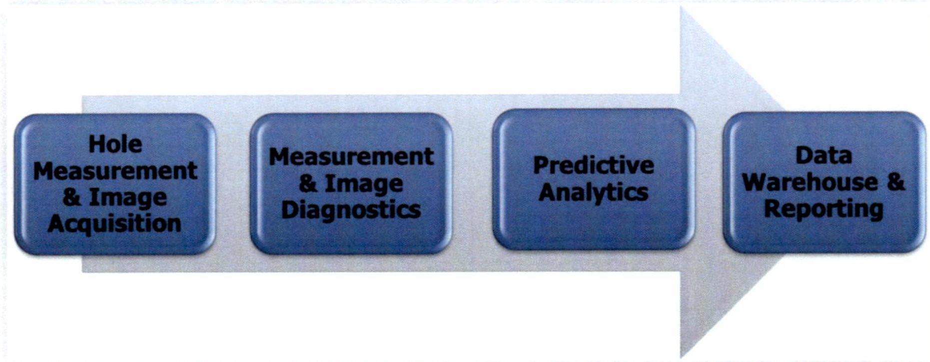

command, communication, and control. The software links automation and handheld devices, providing a complete hole/countersink knowledge-based network across the enterprise value stream.

3.8 Utility

Most handheld gauges for hole/countersink capability continue to grow in complexity, with each additional attribute resulting in training that can cost in excess of a thousand dollars per employee, can last weeks, and requires annual retraining and certification. The ease of operation and functionality of the handheld Aeroscan™ requires a minimum of familiarity for the operator and, therefore, it can transition rapidly into production. The added benefit of a single gauge for the entire hole-bore diagnostics is that the training for operation of one gauge reduces the incidence of error when applying multiple gauges for the same group of measurements. It eliminates human variability of the hole/countersink diagnostics, measurement, and evaluation process in the mechanics' control by performing the operation for them in one simple application. The handheld gauge simplifies the already taxing job of the mechanic and provides a system that complements their skills to achieve better holes/countersinks. The noncontact application of the gauge is more forgiving and compensates for errors in application that result in misreads, missed reads, and rereads of holes and countersinks. In addition, the noncontact operation and rugged construction of the handheld gauge makes it robust and prevents damaging scratches resulting from contact with the airplane skin during its use.

The mechanics or quality personnel have a reduced work burden because the "one gauge" solution is applied once to perform simultaneously the actions of several gauges while simplifying the application action of any one gauge. In addition, the time of measurement and the incidence of error are reduced because of the multiple applications of the various gauges now in operation. By streamlining the application of the hole/countersink evaluation, the intrusion time of the quality control process is reduced so other value-added operations are "freed-up" to enter the workspace and, therefore, throughput is increased.

3.9 Composite Applications

In its infancy, the performance attributes of composite materials for aerospace parts outweighed the complex manufacturing processes and the material, health, safety, and environmental costs as parts slowly replaced metal. The primary focus was on qualifying composite parts onto airplanes and understanding the complexities of replacing well-understood performance and manufacturing processes of metal materials with parts made from the new composite material. The attributes of composites have and will continue to be the primary driver for their increased use on air vehicles.

As more composite parts have made their way onto air vehicles, the design, definition, and macro manufacturing process effects of composite materials have become better understood but remain challenging. As composite material has migrated onto and been integrated into air vehicles, several micro effects within the manufacturing process have begun to manifest themselves and be recognized as contributors to decreased performance. Damage can occur at several scales within the composite material and structural configuration during the drilling operation. This ranges from damage in the matrix and fiber to broken elements, resulting in failure of bolted attachments. The extent of damage impacts load life and residual strength, therefore, it is critical to damage tolerance. Fiber breakage can be critical because structures are typically designed to be fiber dominant (i.e., fibers carry most of the loads).

The assessment of in-hole quality has become a critical component during the composite manufacturing process to ensure air vehicle performance and longevity. The ability to collect, evaluate, and disseminate critical alerts for tool change or process improvement is unique to the Aeroscan™. It stands out as a single solution that can capture and communicate all the necessary measurements for product quality, performance, and longevity. It provides in-process information and data for immediate corrective action based on all the elements of hole-bore quality, including predictive tool changes based on tool wear and in-hole quality.

In an automated application, Aeroscan™ provides a medium for the simultaneous actions of on-assembly hole drilling and evaluation with zero stoppage or interruption. The automated gauging system operates as a coordinated system to continuously provide data at the last hole drilled without loss of time or motion.

The capability to evaluate composite in-hole quality and provide a knowledge base of specific hole-bore quality also facilitates better hand drilling. Specific holes can be identified for manufacturing assessment and refinement to improve hole quality, or mechanics can be identified that may need additional coaching or training where there is not a defect but the hole quality could be enhanced for better product performance.

3.10 Tactical Impact

The improvements in in-processes and results for hole-bore evaluation using the Aeroscan™ system in air vehicle manufacturing processes result in significant savings in the following areas:

- Hidden costs that are included as a part of the manufacturing process
- Design improvements resulting from reduced manufacturing process variability
- Hand drill/countersink process step reductions resulting from the combination of many tasks formerly requiring many gauges into a single task with one gauge
- Automated process step reductions derived from simultaneous drill/countersink and gauging operation
- Hand and automated hole/countersink touch labor reduction
- Enhanced in-process quality
- Above the shop floor support labor reduction
- Cost reductions throughout the supply chain, beginning at the OEM

3.11 Uncollected In-process Costs

There are activities built into and necessary to the drilling, countersinking, and fastener installation processes that are hidden within touch labor cost. One example is grip length measurement for fastener selection. Grip length incorporates safety factors for embedment and thread strip that are important both for the initial installation of the fastener and for long-term reliability with regard to both loosening and fatigue resistance. Grip length assessment for fastener installation was always important but has become more important as composite materials have increasingly made their way onto air vehicles. Composite parts complicate grip length determination because their thickness varies by ±5% and in some parts as much as ±8%.

In metallic assemblies, the mechanic confirms the grip length every 10 or 20 holes because the assembled thickness of metallic parts is very consistent. In assemblies that are composed of stacks of material incorporating composites, almost every hole is checked to determine the proper grip length before the proper fastener is selected. The process of sticking a grip length gauge into each hole to determine the proper fastener grip length is built into either the drilling standards or the fastener installation standards. It is a hidden cost and is time and quality dependent based on the discipline of the mechanic and their visual acuity when reading the gauge. The impact per hole may be only a few seconds when incorporated into the process, but eliminating only two of those seconds per hole when multiplied across one million holes results in 2,000,000 sec, or ~556 h per air vehicle. When producing 600 air vehicles a year, the savings is 333,600 h per year. Using a U.S. $100 per hour rate, savings is U.S. $33,360,000 per year buried in the process, assuming only a 2-sec reduction in process time. This is an unseen cost that can be reduced or eliminated if grip length is calculated at the time the hole is measured.

In the manual grip length assessment, every hole is measured as an independent event based on the composite variability assumption. Data are now not collected to compile patterns in each part geometry to determine areas of repeatable thickness. The mechanic measures the hole, reaches into the appropriate bin in the fastener kit, and installs the fastener. The Aeroscan™ system can collect discrete data and populate a database to perform pattern analysis and map each part for repeatability. This attribute would reduce the variety and the numbers of fasteners stocked at the assembly line and hasten the selection and installation of fasteners.

3.12 Improving Design

It is a little known fact, other than to air vehicle design engineers, that the air vehicle structural design includes performance requirements plus margins for both safety and manufacturing processes. Designers build in safety margins over and above the performance value for a specified air vehicle. They then consider how the airplane will be manufactured and add additional margins by adding such things as material and fasteners to structural components.

Every manufacturing process has variability that manufacturing engineering tries to control through geometric tolerances. If parts are built of composites and vary by ± 5%, a 0.250 in. part can vary from tool location datum plane to trim edge ±0.013 in. If trimmed to a tolerance of +/−0.010 in. and placed into an assembly tool that has been built to a tolerance of +/−0.010 in., and the parts are drilled using drill templates that have holes positioned to +/−0.005 in., then the accumulated tolerance can be +/−0.038 in. Designers have to assume that all the tolerance is accumulated on the negative side and, therefore, add material to compensate for worst-case manufacturing outcomes. When automated drilling increased the diametric positional accuracy for holes and used optical positioning for best edge distance fit, the designers removed material from the substructure knowing that the new manufacturing process would reduce part variance. They were conservative and only reduced the trim-to-datum thickness by 0.005 in., but this equated to miles of linear inches of material removal in substructure across the airframe. The material removal saved hundreds of pounds of airframe weight without sacrificing performance or safety margins.

Hole-bore quality can be predicted and improved so that design engineers are able to evaluate and refine designs that reflect a manufacturing process that has less variability and is closer to the design intent. **Figure 3.4** shows use of the as-manufactured Lower Control Limit (LCL) Design Engineer as consideration for design allowables.

FIGURE 3.4 Use of as-manufactured Lower Control Limit (LCL) Design Engineer as consideration for design allowables.

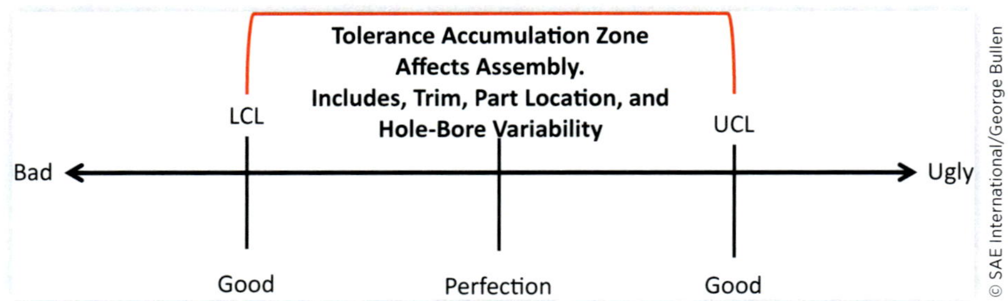

3.13 Explanation of Automated Gauge Operation

Integrating a gauging application/solution aside and functioning as integral to hole-bore preparation (drill and countersink) is essential for efficient automation. The automation of drill and countersinking, combined with fastener installation, is separated by the interim step of multiple gauging operations necessary to validate the quality of the previously drilled and countersunk hole prior to fastener installation. The complex array and quantity of gauges necessary for all the hole-bore sizes and types of measurements necessary to ensure the hole is properly prepared for installation of a fastener complicates automation. The chasm created by the complexity of the gauging process defeats closure of the two primary components of automating the drill and fill process: drill and countersink and installation of a fastener. Further complicating the closure of these two disparate operations is the added need for enhanced hole-bore and countersink evaluation when composite material is added as a material component of the parts to be joined. The enhanced hole-bore and countersink evaluation often involves "eyes on" the hole to use human subjective evaluation to ascertain the quality of the hole. This is done because current gauges meant for an isotropic material (metal) are not capable of properly evaluating holes drilled and countersunk in composite materials. Combining multiple gauges into a single gauge and the capability to evaluate the quality requirements of holes drilled through composite material make this type of gauge the solution for a highly capable and efficient automated drill and fill operation. The design of the gauge presents a form factor that is easily integrated into and between the drilling operation and the fastener fill operation.

3.13.1 The Basics

The basic principle and essential process that is one of the primary enabling technologies for drilling on-assembly is the ability for the automated drilling machine to "find" the skin of the airplane work-piece with a high degree of precision. The components of a work-piece assembly are often composed of various types of material and are held in place with a variety of fiduciary fastening devices, often in an environment that is uncontrolled (temperature and humidity). The open work environment, means of clamping, and the thickness variability of composite skin combine to exacerbate the location of the key component for attaining the precision required to drill and countersink holes in aerospace assemblies. The skin Outer Mold Line (OML) is the key to success for depth control of the countersink. It is the beginning datum that controls the drill algorithms essential to changing speeds and feeds as the drill progresses through the different stacks and types of materials. And this key measurement determines the drill tip exit point where the drill is extracted. If a hole-bore is drilled too deep and extends too far beyond the minimum required depth to complete the hole, then the drill tip can impact tubes, wiring, or other components that exist behind the substructure. Controlling this measurement prevents drill starts, a term commonly used in the aerospace industry to identify defects created by an unintended impact by a drill bit tip with a critical component. Of course, the machine is positioned and the integrated design of the machine, parts, and tooling is constructed so that they all closely approximate the OML of the skin when combined. The construction and coordination of machine, parts, and tooling are close, but do not rise to the level of precision needed.

To solve this technical problem, almost all[1] on-assembly automated drilling machines use a pressure foot that is attached to the drill motor and is part of an automation component called an end effector. The pressure foot is located at the front of the end effector and, therefore, is the first material component of the end effector to "touch" the skin. The pressure foot is also attached to a precision linear scale. The end effector and pressure foot "face" are moved toward the skin of the airplane by means of a precision linear actuator. A drill bit tip with a countersink component is precisely located in relationship to the "face" of the pressure foot. **Figure 3.5** is a notional image of the construction of the end effector and its relationship to the skin.

FIGURE 3.5 Notional image of the construction of the end effector and its relationship to the skin.

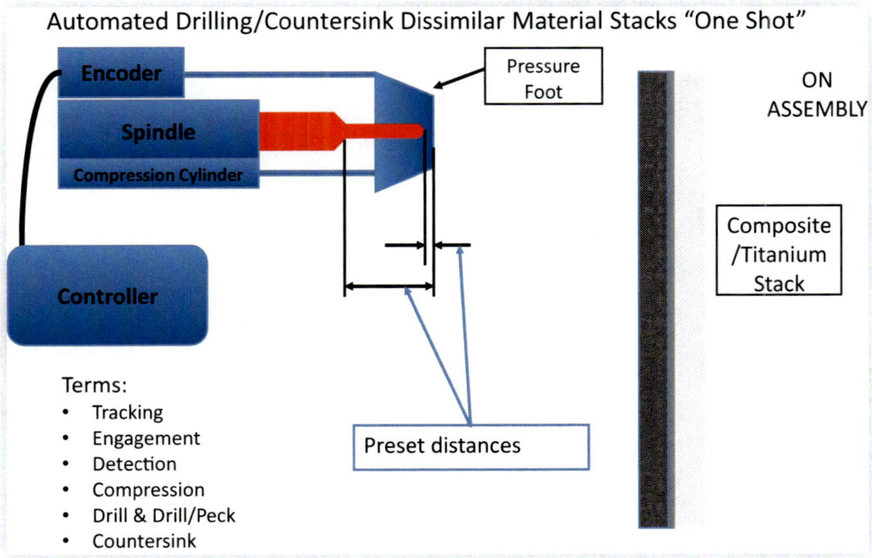

[1] I use the term "almost all" because I assume I have not seen every on-assembly automated drilling machine that exists. And I can imagine that there could be instances where under certain conditions the described essential technology would not be needed.

When the on-assembly automated machine has moved to a position to drill and countersink the hole in the work-piece, a command is sent to the end-effector linear actuator to move slowly toward the skin. When the pressure foot touches the skin, the attached linear scale is compressed notifying the controller software to begin the drill and countersink routine for that specific hole. The hole-bore specifications, material types, stack thicknesses, and other critical information are derived through association with the machine position for the hole to be drilled.

3.13.2 Gauge Integration and Operation

All the technologies designed and refined for the drilling process are complementary for integration of an Aeroscan™ type gauge into the on-assembly automated drilling and countersinking operation. In its first application, the gauge was positioned in between the drill motor and the fastener installation mechanism. Figure 3.6 shows the gauge positioned underneath the pressure foot after the drill motor portion of the end effector has moved to the left. Since the pressure foot remains compressed against the skin, the reference datum for acquiring the beginning gauge measurements (countersink depth and grip length) is established and coordinated with the just-completed drill/countersink operation.

FIGURE 3.6 Gauge positioned underneath the pressure foot after the drill motor portion of the end effector has moved to the left.

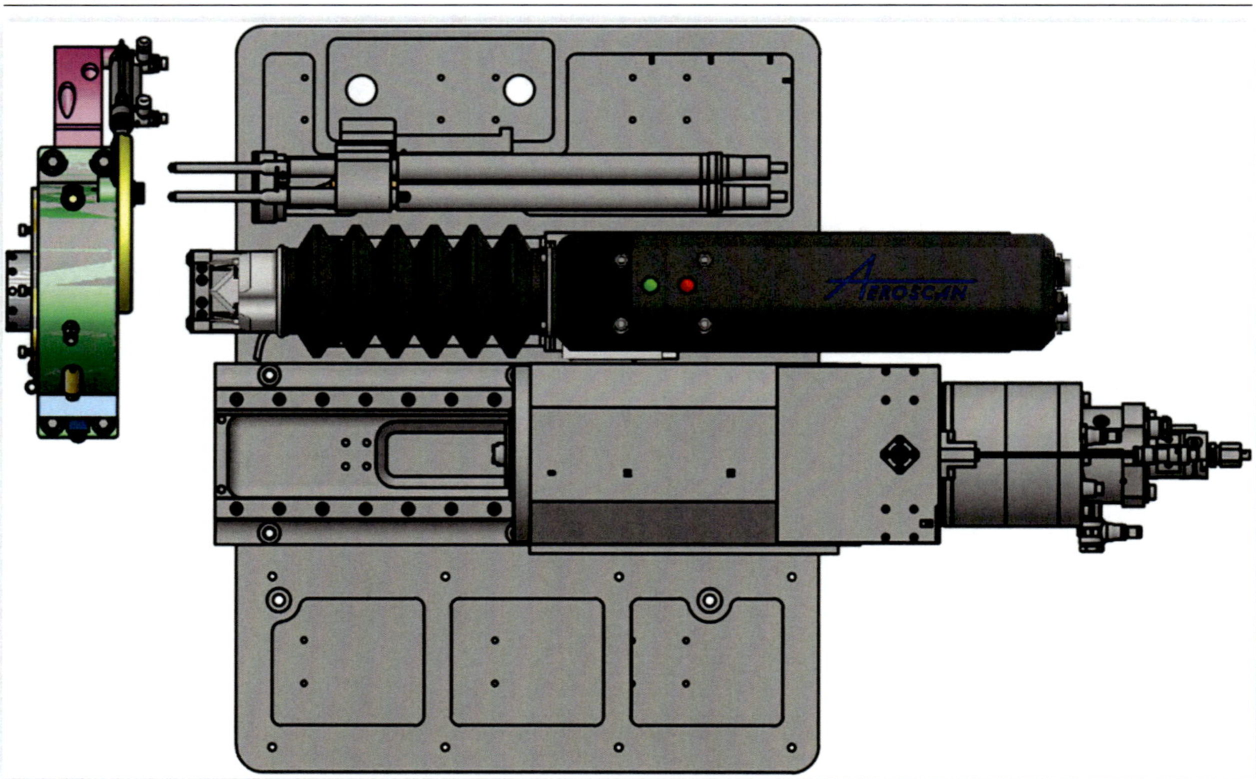

FIGURE 3.7 Gauge as it has moved up into the pressure foot.

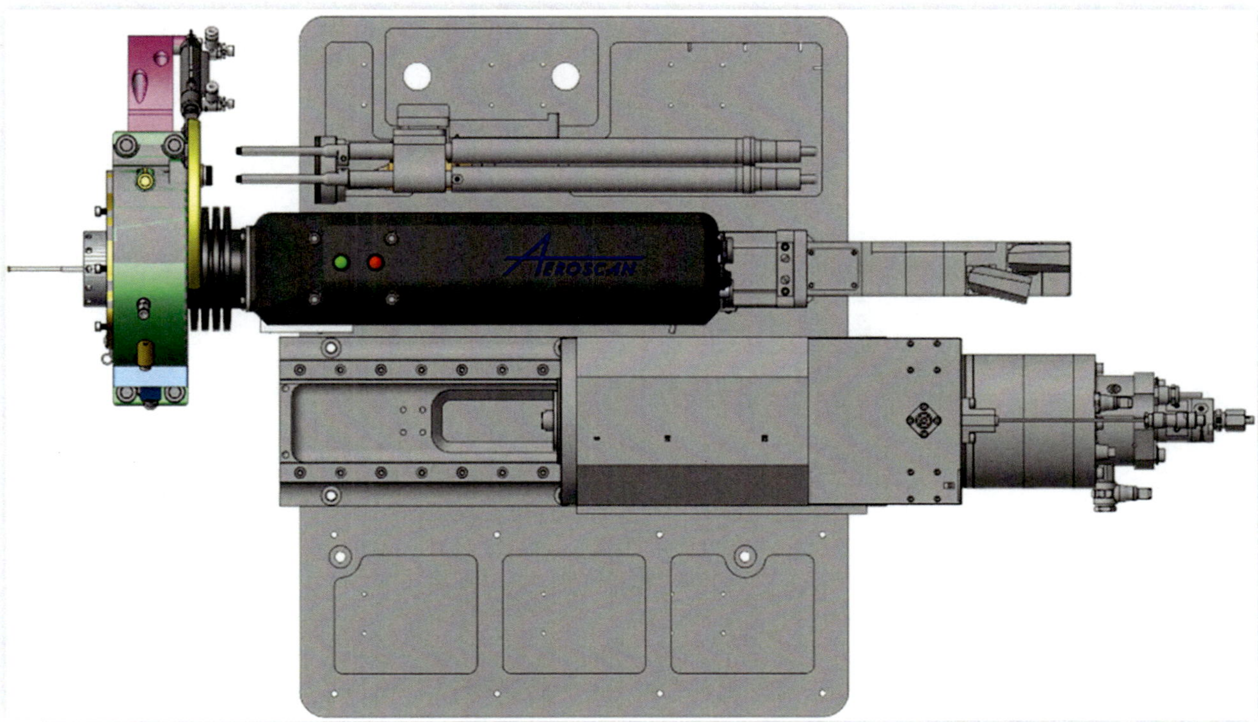

Figure 3.7 shows the gauge as it has moved up into the pressure foot. **Figure 3.8** is a close-up image showing the gauge in the pressure foot before retraction.

After measurements are taken, the as-measured data are compared with the requirements for all critical specifications. If the criteria are met, the machine moves to the fastener installation position and the fastener installed. If the criteria are not met, the machine moves to the next hole and the drill countersinking routine starts from the beginning. In all cases, the user interface displays only the go/no-go analysis with an image of the critical measurements. **Figure 3.9** shows the data displayed on the Human Machine Interface (HMI) to the operator.

FIGURE 3.8 Close-up image showing the gauge ir the pressure foot before retraction.

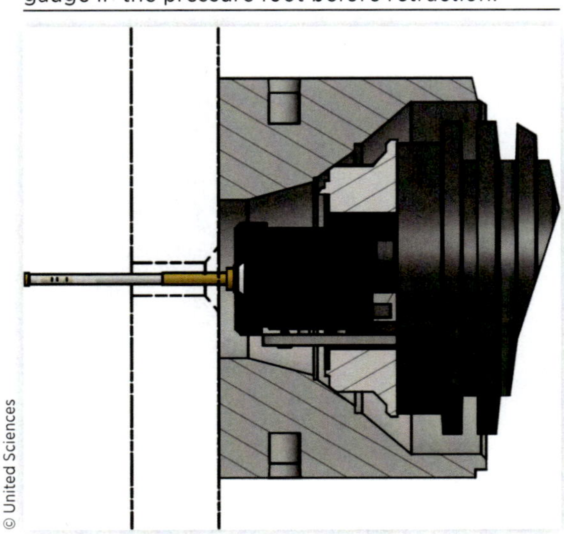

Other data are stored and used for advanced analytics to compare and adjust future operations to continuously improve the quality of the hoe and its attributes. Hole-bore imaging, borehole fingerprinting, and data storage provide important analytics and diagnostics for the life of the plane (may be shared between manufacturer, maintenance, repair, and overhaul for optimized performance and failure analysis). Drill life assessment such as predictive analytics to assess drill life and identify a precise replacement interval can be determined. **Figure 3.10** shows the damage that normally would not be detected by current gauges. The drill bit has begun to tear, rather than cut the surface revealing the need to replace the drill bit.

FIGURE 3.9 Data displayed on the Human Machine Interface (HMI) to the operator.

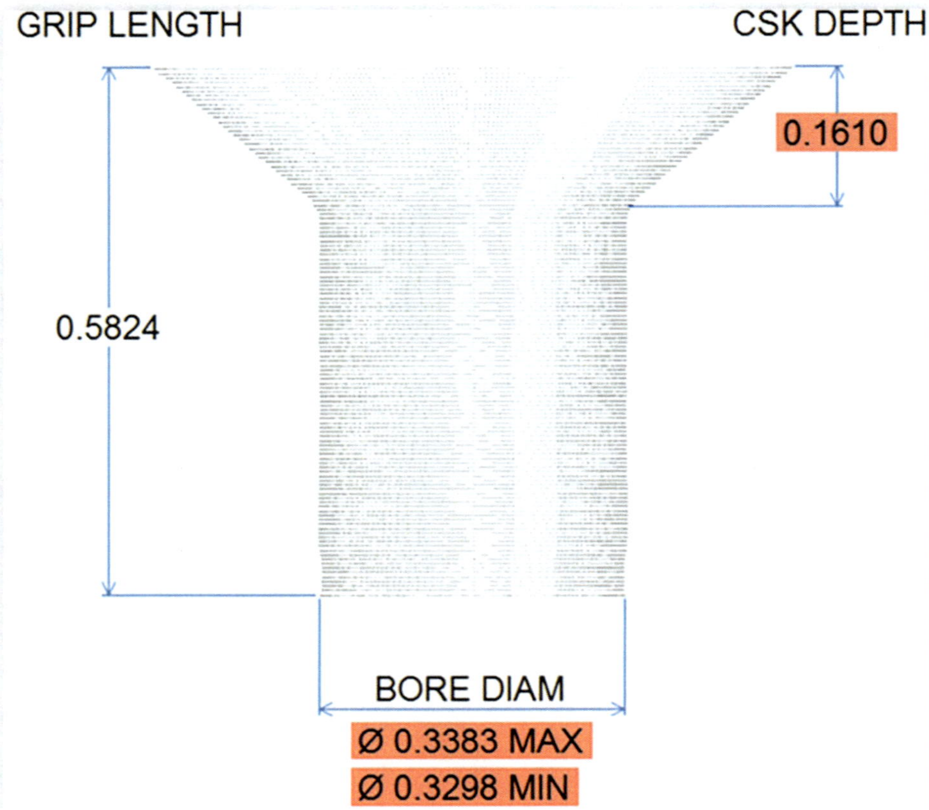

FIGURE 3.10 Damage that normally would not be detected by current gauges.

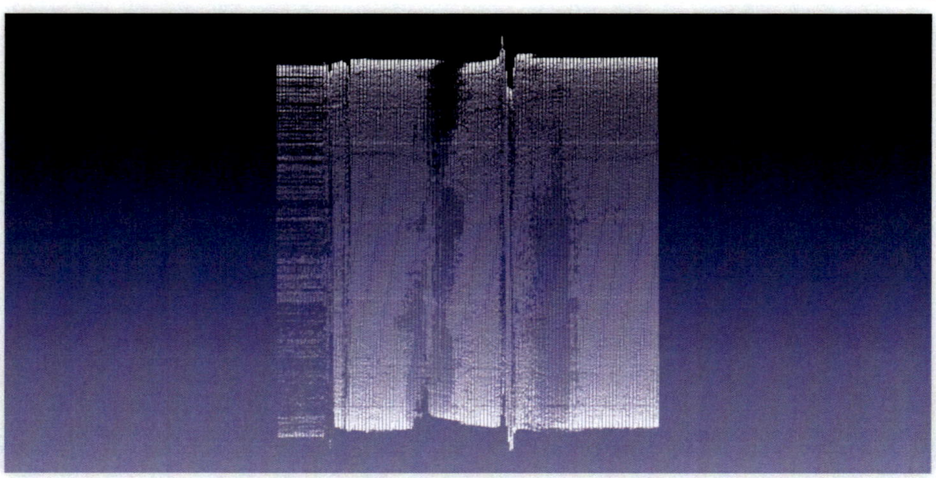

Figure 3.11 shows an anomaly called a "snow man" in aerospace vernacular that was passed by a gauge meant for metal but discovered by a ring-laser optical gauge (Aeroscan™) (image courtesy of United Sciences).

The logic for moving to the next hole if the hole is not acceptable is to keep the expensive piece of machinery moving and fully utilized. The hole can be evaluated for cause and corrective action by hand methods as the machine continues to drill and fill holes. The exception to this operational rule is if two consecutive holes show defects. This may indicate a machine maintenance issue, undetected drill bit damage, or operating deficiency that may have to be addressed and corrected.

Almost every end effector must have a pressure foot or other means to identify the OML of the skin prior to beginning the drill and countersinking operation. Variations in end-effector manufacturer designs can change the gauging application just described. For example, the KUKA system uses a rotary system in which the pressure foot disengages after drilling/countersinking and the end effector rotates to the next position like the operation of a Gatling gun. Each position contains its own pressure foot for acquisition of the skin location. The process for moving toward the skin and compressing a linear actuator is performed at every operation. The drilling, gauging, and fastener installation modules are self-contained in tubes that are removable for repair, maintenance, or refurbishment.

FIGURE 3.11 An anomaly called a "snow man" in aerospace vernacular that was passed by a gauge meant for metal but caught by a ring-laser optical gauge.

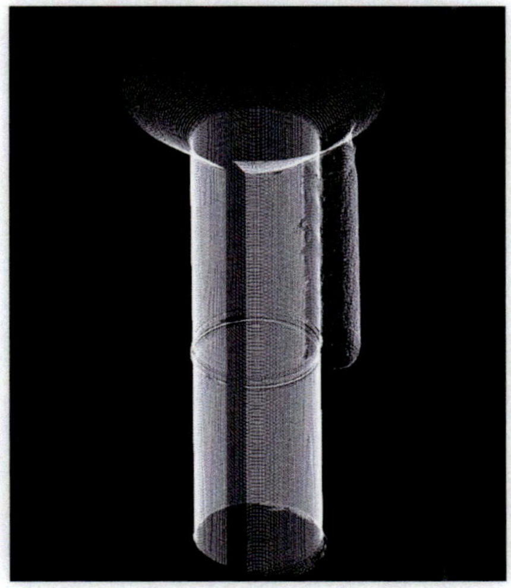

© United Sciences

Other manufacturers mount the gauge to the side of the drill and countersink end effector where the drill countersinking operation is performed and then the machine moves to position the gauge at the centerline of the hole-bore. An independent linear actuator for the gauge then moves the gauge into and out of the hole to perform the gauging operation.

3.14 Explanation of Handheld Gauge Operation

Taking a precision gauge that must be properly positioned and aligned with the centerline of hole-bore and holding it in place manually creates design challenges that dwarf adapting the gauge to a machine. There are several types of handheld applications that can be applied. The easiest is to use a purpose-designed nosepiece that is the same size as the power feed motors used in the aerospace and automotive industry today. The templates that are used for drilling (holding the power feed motors during the drilling operation) are then also used to accommodate the gauge nosepiece where the operator installs the gauge into the bushing after the hole has been drilled. The gauge is activated by pushing a button and the gauge takes measurements without further human intervention. The gauge is removed after the measurements are taken and moved to the next hole. Insertion of the gauge into a hardened steel bushing that is bonded or press fit into a metal or thick laminated drill template holds the gauge precisely and solidly in place during the gauging operation. The human contribution is to remove the gauge from the hole-bore just measured, move it to the next position, and insert it into the bushing of the drill bonnet for the next hole to be measured.

FIGURE 3.12 The original design for a freehand noncontact handheld gauge solution.

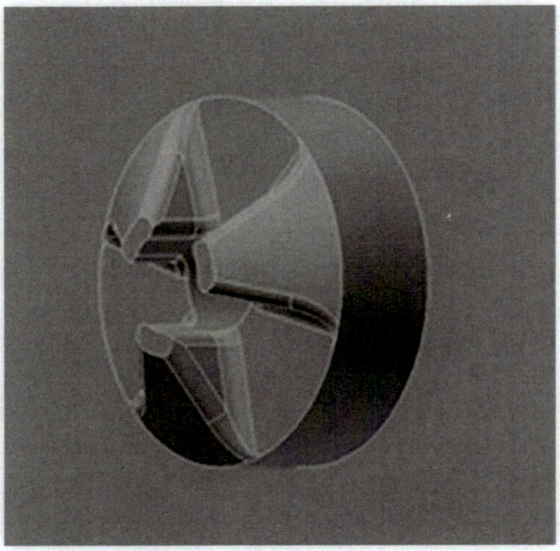

© United Sciences

A greater design challenge for a noncontact gauge like Aeroscan™ is to design in a "fail safe" method for precisely positioning and having a human hold the gauge in position during the seconds that it would take to gauge a hole-bore and its countersink. This free-hand gauging method presents a unique challenge that originally led to a design with a conical nosepiece. Figure 3.12 shows the original nosepiece design for a freehand noncontact handheld gauge solution.

The nosepiece was meant to align the gauge with the hole-bore centerline and three equally spaced micro-switches on the nosepiece would activate the gauge when all three were in contact with the skin. Once all three micro-switches were in contact, a light would notify the operator when the gauge was normal to the surface of the part and they would activate the gauging sequence.

This solution is not tenable over the many holes to be gauged during an individual's work shift. In the short term and in a lab environment, it showed promise but could not be replicated in operation in a shop environment, an environment in which the operator had to repeat the operation thousands of times with noise and distraction a constant.

Therefore, solutions have been derived from different gauge operators such as Boeing and Airbus, who are airplane OEMs where the value of the gauge has incentivized them to take on their own developmental efforts. The goal to design a "free" handheld gauge, specifically for their applications and use that is robust and repeatable in the dynamics of a shop environment.

3.15 The De-stack Solution

It is of great amazement to people unfamiliar to aerospace and, in some cases, automotive assembly process that a manufacturer would tightly control all the steps necessary to fasten an assembly together and then instead of fastening it together.........take it all apart and then put it back together.

There are many reasons for this seemingly unnecessary process. Most important are the hidden effects of drilling through two pieces of material at the interface of the two materials to be joined. Compounding the manufacturer's concerns was the introduction of composite material as the top or outer piece of material to be drilled. This necessitated pulling a piece of (most often) metal material through the softer outer material that is increasingly becoming the material of choice for skins of cars and airplanes. Pulling a harder piece of material through a softer piece of material has obvious disadvantages. The hole-bores in aerospace and increasingly in automotive are tightly controlled with very tight tolerances including diameter, concentricity, and uniformity. Uniformity refers to the entry of the hole-bore to be the same or very nearly the same diameter as the exit diameter. Pulling an aluminum, steel, or titanium chip with a sharp knife edge up and out through a carbon fiber hole requires very controlled and nuanced adjustments to get an acceptable hole-bore front-to-back. And the process

FIGURE 3.13 The cost of de-stacking an airplane subassembly.

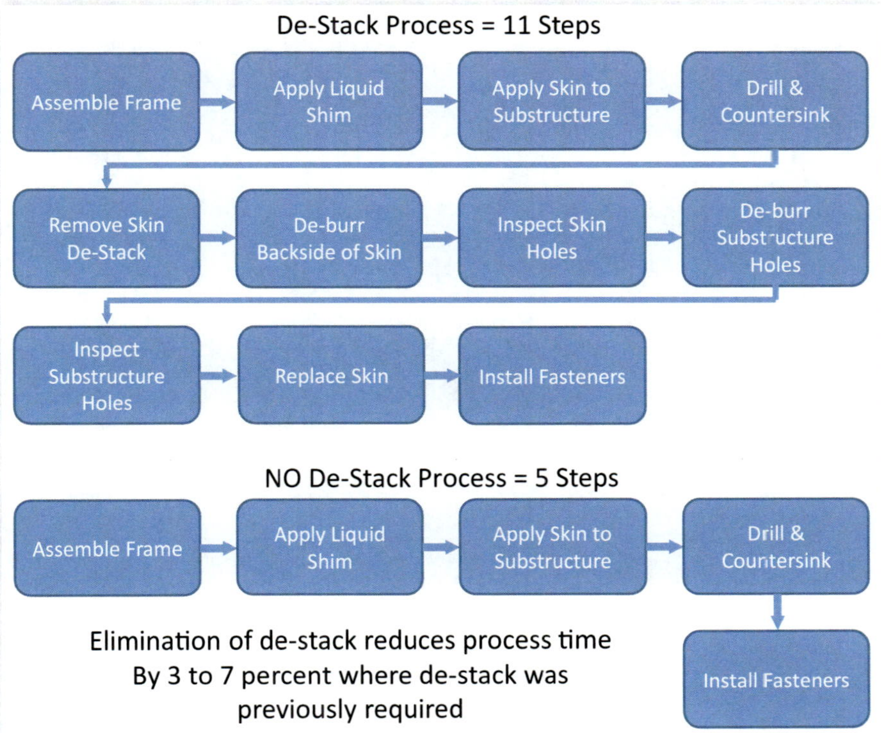

is made more difficult by the efficiency requirement of eliminating step drilling in favor of drilling a finished hole "one shot" front to back with one drill in a single pass.

Complicating the delicate process was the effect of drilling through dissimilar materials at the interface of the two pieces of material. Separation, burrs, debris, damage, and other complications could not be ascertained by either the current gauging methods meant for metal or by a person looking into the hole. So, once drilled, the skin had to be removed from the substructure so that the effects of drilling could be properly evaluated at the interface between skin and substructure. Figure 3.13 illustrates the steps and cost of de-stacking an airplane subassembly and why aerospace is vigorously pursuing a way to illuminate the process.

The fidelity of a noncontact gauge using ring laser technology gives a complete (holistic) image of the health of a hole, including the interface between the two or three pieces of material.

When this type of fidelity is provided, there is no need to de-stack an airplane. An anomaly derived from the drilling problem can be identified, addressed, and corrected while the assembly is joined. In many tests, the incidence of interface defects was so rare (1 in 1000) that de-stacking the assembly for a 1 in 1000 occurrence was prohibitively expensive. However, the defects that are discovered even if one in a thousand can cause failure or in some cases catastrophic failure so that in the absence of a means to see the defect, de-stacking the assembly is the only alternative.

Using a ring laser noncontact type gauge, these types of defects can be discovered while the airframe is joined and methods have been developed to mitigate their presence and so there is no need to de-stack the airplane. Figures 3.14–3.16 show burrs that the Aeroscan™ type gauge discovered that would otherwise been undetectable without de-stacking the

FIGURE 3.14 Burr at interface looking in from the side.

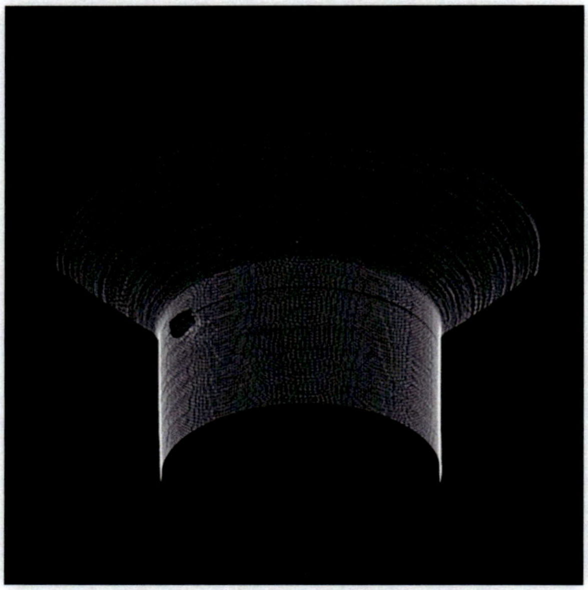

© United Sciences

FIGURE 3.15 Burr at interface.

© United Sciences

FIGURE 3.16 Burr at interface defined and sized for removal.

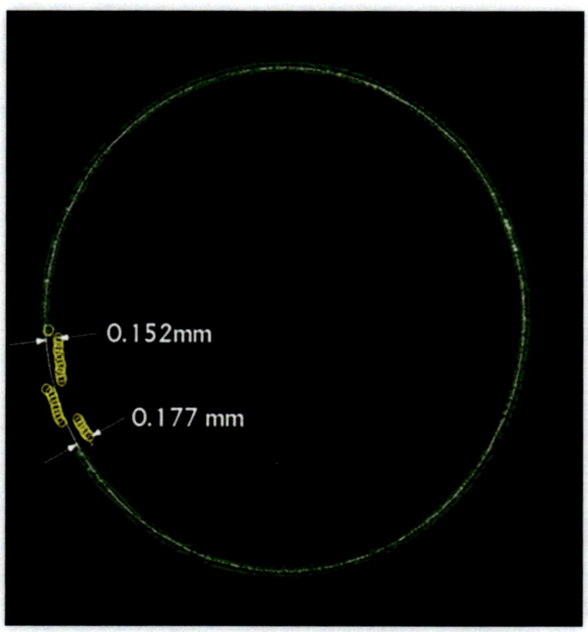

© United Sciences

assembly. Using this type of gauge can also determine the size and precise position of an anomaly defect such as burrs so that they can be accurately corrected.

This type of technology also provides the visibility to see all other defects that the de-stacking process is meant to expose.

3.16 Summary

The operation of this type of gauge seems to be the answer to many issues affecting the cost of quality and efficient assembly of airframes. Also, it has passed the rigid testing for application and migration across manufacturing at many companies including major OEMs like Airbus.

The adaptability of the form factor is easily integrated into the automated operating systems of most of the primary end-effector manufacturers for on-assembly drilling and countersinking. Its operation is straightforward and its output dwarfs the capabilities of current gauges.

So why the delay for its transition to production? There are many reasons.

The replacement of hundreds of thousands, if not millions, of gauges that have become an accepted means of validating the primary source of holding everything together is no small task. Vehicles that carry people and especially air vehicles are complex machines and are becoming more complex every day. Planes fly themselves and cars are beginning to drive themselves. All the processes that hold them together are validated by a comprehensive set of processes using traditional and clearly understood gauging devices.

New materials have begun to drive a new look at how we evaluate the joining process, and the gauges we use have come under increasing scrutiny as not having the capability to assess and evaluate the new materials such as composite materials that are migrating onto our vehicles of today.

The conundrum is validation of a gauge type that clearly offers a single gauge solution, thereby simplifying the validation process. Validating a gauge that also offers a healthy reduction in cost by elimination of de-stack and increasing the opportunity for automation takes time. Replacing the known and accepted in such large numbers cannot happen overnight.

Furthermore, the complicating migration of a one-gauge solution into the assembly operations of many manufacturers is a concept of checks and balances. Current quality processes use a variety of gauges that include the human eye. Make a mistake with a countersink gauge and use all the other gauges correctly and the chances are good that the hole-bore, grip length, and other attributes that contribute to fastener strength will compensate for the oversight.

The ring-laser noncontact gauge does not touch the hole and provides a one-measurement solution, including elimination of the "eyes on" security of validation. One gauge without human "sight" for one final check is frightening to those who manage risk.

That said, the ease of operation and capabilities to fully gauge all aspects of a hole, including those needed for new materials, will drive this technology into and across the air and land transportation manufacturing industrial base.

CHAPTER 4

Application

Things work out best for those who make the best of how things work.

—John Wooden

4.1 Introduction

The introduction and application of any new process that disrupts the rhythm and flow of a decades-old proven and robust manufacturing process must be evaluated, carefully planned, and transitioned into the manufacturing flow. Meticulous calculations must be performed to minimize the trauma of change to the people, technology, and processes that now perform the accepted and understood process using devices the new technology will replace. No matter what the value in savings and quality the new technology or process will provide, even the simplest change in a manufacturing process or technology that has matured, become habit, and accepted can have catastrophic results. On March 7, 2014, Boeing Company said that "hairline cracks" had been discovered in the wings of about forty (40) 787 Dreamliners. The 787 wing-maker Mitsubishi Heavy Industries Limited notified Boeing in February of the problem, which arose after the Japanese company altered its manufacturing process [4-1]. Mitsubishi's minor and unreported change to the wing skin manufacturing process is only one example of a seemingly insignificant change impacting a product's cost, safety, and delivery in a major way. "If it is not broken, don't fix it" is not just a phrase in the manufacture of products. The highly complex manufacturing processes of airplanes and automobiles make this phrase a law rather than a rule.

Aerospace and, to a lesser extent, automotive manufacturers are very conservative when it comes to disruptive change to their manufacturing processes or the technology they use for good reason. The products they build are extremely complex and dependent on the manufacturing processes they use to be repeatable and in control. The consistency, reliability, and safety of their products demand it. They are also composed of complex manufacturing supply bases spread across the globe. In order for all the pieces of the product to come together at final assembly centers and fit precisely, common processes and technologies must be clearly understood, controlled, and used. They know how to build airplanes and automobiles, and the risk to change the current process must have very compelling reasons.

In my book *Automated Drilling and Countersinking of Airframes* [4-2], I describe a precise process that was used and continues to be used to transition new technology into production while minimizing the impact to the currently used, understood, and embedded technologies and processes. The highly complex world of on-assembly drilling and countersinking has "taken hold" and migrated across the production landscape until the application of the technology and its associated processes have become almost commonplace. But it was a two-decade long journey, fraught with failure by those who did not follow the necessary and defined process for transitioning it to production.

4.2 Familiarity—Overcoming the Norm

Combining traditional handheld contact gauges used for decades and accepted as the only method for determining critical values in the manufacturing process into a single noncontact gauge and transitioning it into production is daunting. The current gauges are feel, touch, and see-type measurement gauges, where humans can feel the gauge interact with the surface to be measured, touch or hold the gauge in their hand as it touches the surface, and see the gauge interact with the surface as it displays measurement values. There is comfort and security in touching and feeling to human security and understanding. Rick Chillot's article *The Power of Touch* in Psychology Today [4-3] points out that, "Touch is the first sense we acquire..." and researchers, such as DePauw University psychologist Matthew Hertenstein demonstrated that we have an innate ability to decode emotions via touch alone. In fact, the critical component necessary for the success of every on-assembly automated drilling and countersinking system used today replicates the human-touch sensory feedback system to determine when the machine is in contact with the surface of the work-piece. It is only when the contact "sensation" is communicated to the machine controller that the drill/countersink algorithm is activated. We as engineers like to think that data alone validate and give comfort to our lives. But the fact remains that the critical nature of the measurements taken by the prevalent touch gauges used by mechanics on the assembly line provides the necessary confirmation that, "everything is in control." Mechanics know how to use the gauges and are comfortable with their use, they provide the necessary data, and are historically proven to validate critical processes.

The contact area of the gauges used now that touch the work-piece is also steel. They are rugged and proven. The shop floor is not a gentile environment.

The point is that the introduction and application of any technology to replace the current collage of measurement gauges needs to replicate the current process as closely as possible. And it must be robust.

There are examples of non-touch systems in use on the factory floor of aerospace and automotive industries. Leica and Faro (among others) laser systems provide line-of-sight laser measurement systems to validate dimensions on parts, assemblies, and fixtures. They do require a target affixed to or held against a part, assembly, or tool. They are single-point measurement systems that they measure fixed targets or targets held by a human. Therefore, there is an association with the see-and-touch component that provides the comfort level for human acceptability.

The current batch of gauges does have issues that are accepted as a necessary part of their operation. Many gauges must be used to perform all required measurements across the manufacturing landscape. One gauge cannot measure across all hole-bore sizes inherent in airplane or automotive product manufacture. Nor can one gauge measure all the countersink angles and sizes while providing grip length information. As previously mentioned, the current gauges were developed for specific application on metals and are inadequate to give the added data points with the fidelity required for validating composites.

The volume of gauges, designed for specific attributes (purpose-built), and failure to address the unique issues for gauging composites have challenged on-assembly machine integrators to automate the gauging process. Many gauges contained in a tool changer that have to be retrieved every time a hole is drilled and countersunk slows the process time and defeats a positive return on investment to replace the hand method of gauging. When the gauging process takes longer than the drill, countersink, and fill operation, then there opens opportunities for new technology acceptance.

4.3 How Much Is Too Much?

Another component for acceptability and application of a noncontact, one-size-fits-all, ring laser gauge is the large amount of data that it collects and displays. Production and process engineers are often at odds about what data and how much data are needed to provide adequate assurance that all is as it should be. Product delivery and manufacturing because of its critical-time-based environment only wants to know what is expected to validate a measurement in the simplest form. Is it good? Or is it bad? And they want the production mechanics to know only two things when they evaluate (measure) with the current batch of gauges. Is it good? Or is it bad?

If good-to-go, then move on and continue the product flow to completion. If it is a bad measurement, tag it and let the engineers figure it out and recommend a fix (cause and corrective action). Keep moving is the mantra. In a perfect world, continuous and large amounts of data would be good to "see" every aspect of the production process at the micro and macro levels. The data would be used to enable advanced analytical software to continuously evaluate, adjust, communicate, and relate data for best outcomes. The data would be used to facilitate independent actionable machine decisions for the best fit at each level of assembly. This is scary stuff at the shop level where life has been simplified to, "Did I meet my quality, cost, and schedule goals?"

Big data scares production and manufacturing operations because it complicates a simple go/no-go approach to production happiness. When coordinate measurement machines (CMM) began migrating to the quality world, they offered to provide a level of measurement accuracy out to 0.00001 in. Parts formerly deemed acceptable by traditional gauges were now rejected as being 0.00001 in. out of tolerance. To put this minute dimension into perspective, human hair is on average 0.004-in. (0.1 mm) thick.

This was a very real problem that continues today. When I was a tooling manager on the B2, a tool my team had built to fabricate the toilet roll enclosure for that airplane was rejected for having one of its radii out of tolerance to 0.0002 in. (two ten-thousands of an inch). The newly acquired CMM had rejected the tool because it had the capability to look beyond what would be called reasonable consideration. When I argued my case to the quality manager (adding in the noncritical nature of the part), he responded that the reject was now a matter of record. It was outside of acceptable tolerance and, therefore, his ability/authority to override the rejection. The tooling team had to remake the tool. Not to be dramatic, but I must reiterate that this was for a toilet roll holder.

This big data fear is pervasive in production operations and creates a conundrum for aerospace manufacturers. They want and in the case of composite parts need more data. They want more data because they intellectually recognize the benefit to the product's quality, cost, and schedule derived from the best fit at assembly and in-process systems health monitoring. They understand that computers can rapidly analyze and streamline communications and enhance assembly processes while lowering cost and improving quality. But they want all that hidden from the production floor to maintain the simplest processes and the most straightforward prompts and queues necessary to speed parts through assembly.

So one of the application barriers for a noncontact ring laser system that provides a 40,000 point data cloud is to clearly establish what will be used at the shop floor level to maintain the required simplicity necessary for fluid production operations, and to leverage the same information (behind the scenes) through advanced analytics to enhance the best part and assembly outcomes. All while providing (at a minimum) the same level of confidence and fidelity of the current gauges expanded to include the new requirements for composite materials.

A cornerstone for inserting any new disruptive technology[1] into long-established processes is to minimize disturbance to the current process. Transition-to-production (TTP) teams are tasked to try and place new technology into the current process to enhance it while reducing any changes needed to incorporate it. Adherence to this basic principle has led to success, while ignoring it has led to failure. To smoothly insert a new technology into an existing process, it is important to understand all the elements of the current process.

[1] In aerospace, government research, and universities the term "disruptive technology" can be used positively or negatively. In this case, it is used positively to describe a technology so revolutionary that it positively impacts cost, quality, manufacturing flexibility, and schedule.

4.4 Current Processes

Air vehicles in use nowadays are drilled and countersunk in assembly operations using three primary methods. They are hand, mechanized, and automated drilling and countersinking.

4.4.1 Hand Drill Hole-Bores Using an Air Operated Drill Motor with Multiple Drill Bits in Progressive Steps, Countersink, Clean, and Measure

A handheld slip renewable bushing is installed into standard-sized holes in a thin aluminum template affixed to the work-piece and positions the drill bit in reference to the work-piece. The drill template is attached to the work-piece using precise holes in the template that align with prepositioned holes in the work-piece. The predrilled holes in the template are aligned with the tool reference system that is in turn aligned with the engineering coordinate reference system. The template is held in place by Cleco fasteners. Cleco fasteners are temporary fasteners that precisely align and hold the template tight against the work-piece. They are either spring loaded and installed by means of special Cleco pliers, or they are tightened using a threaded nut by means of a special pneumatic handheld motor. The type of Cleco to be used is dependent on the size and weight of the template. Spring-loaded Clecos are used for smaller lighter templates, and screw type Clecos are used for heavier larger size templates.

The hole in the work-piece is drilled in progressively larger steps beginning with a small pilot hole until the finished size is reached. The template is then removed, the work-piece surface thoroughly cleaned, and the countersink operation performed using a micro-stop countersink cage with the proper angle, piloted countersink installed, and the cage adjusted to the proper depth. All of these operations are performed by highly trained and skilled mechanics, since the position of the holes and the finished size of the hole-bore and countersink depth are held to very close tolerance. Hole-bores are commonly held to tolerances of the hole-bore diameter of −0.000 in. and cannot exceed 0.0015 in. beyond the hole-bore diameter. An example is a 0.250-in. hole where the acceptable range for a finished hole-bore would be 0.250–0.2515 in. Countersink depths are held to varying tolerances, depending on their location on the airframe and the function of the air vehicle. Generally, fixed wing military air vehicles have closer tolerance countersink depth control criteria than commercial fixed wing air vehicles. Generally, rotary wing aircraft have looser depth countersink depth control criteria than fixed wing vehicles. Typical depth control can range from +0.002 in. above the surface the fastener is to be installed to 0.010 in. below the surface.

The hole-bore must be drilled perpendicular to the surface it penetrates and have good material finish. In aerospace vernacular, a hole-bore is "normal" to the surface if it is perpendicular to the entry-point surface of the drill bit. Figure 4.1 shows the tangent/normal relationship. The drilling of

FIGURE 4.1 Tangent line with normal to surface entry point of the drill bit.

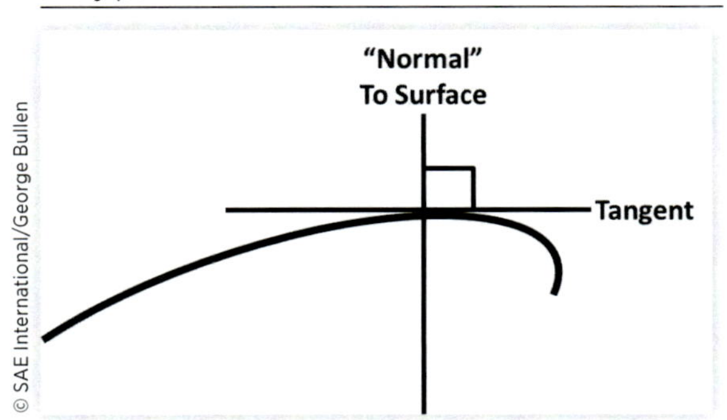

FIGURE 4.2 Illustration of a gapped fastener.

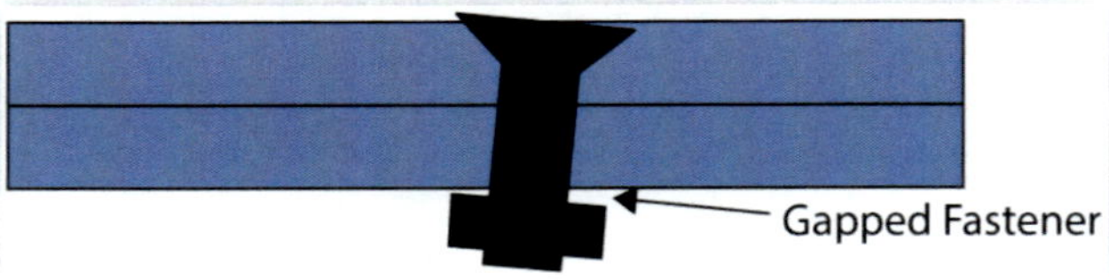

the hole-bore is a critical component in the drilling process. If the hole is not drilled normal to the surface, then the fastener does not seat properly into the countersink side and the nut on the backside of the fastener will have a gap. This condition stresses the fastener and reduces its strength. **Figure 4.2** shows a hole-bore out of normal resulting in a gapped fastener.

All of the drill and countersink dimensional components previously described are critical to the viability and performance of the fastener.

The backside (where drill bit exited) is de-burred after the finished hole-bore size is drilled.

The work-piece surface and hole-bore are cleaned after the drill and countersinking operation.

It should be noted that in many instances in which the material stacks to be joined are aluminum over aluminum, a de-stacking operation must be performed after the drill and countersinking operation. The previously clamped together pieces are taken apart (de-stacked) and inspected. The operation is done to ensure that the mated surfaces the drill has passed through have not been damaged, collected debris, or incurred burring. Almost all instances where the pieces to be joined are composed of at least one stack material-type of composites must be de-stacked after drilling. Additional complications and repairs may occur where composite materials are a stack component because of the shimming materials added between the two materials joined together. The shimming materials are added to remove gaps inherent in composite materials rough side (bag side) surface that must be filled to provide a tight fit between the two parts joined together. While the liquid shim material fills the gaps between the two joined pieces, the brittleness of the material when cured adds complexity to the drilling process. Cracking of the liquid shim material may result from the drilling operation that can only be determined and repaired (if present) by de-stacking the pieces and doing a visual inspection.

Inspection dimensional gauge points occur after the holes are drilled full size, after countersinks are bored full depth, and before fasteners are installed. Prior to fastener installation, a gauge is inserted into each hole-bore and the length of the hole from the countersink surface to the exit or end of the hole-bore is measured. This measurement determines the length of the fastener to be installed in the hole. This is called a grip length measurement and is critical to proper fastener selection.

4.4.2 Mechanized Drilling Using Power Feed Motors to Drill Hole-Bores with Multiple Drill Bits in Progressive Steps, Countersink, Clean, and Measure

The mechanized process for drilling and countersinking holes varies very little from the handheld process described in (A), except for the use of power feed motors locked into hardened steel bushings located in the drill bonnets. And the drill bonnet is a larger, thicker, and more robust medium for aligning and positioning drill bits to work-piece. **Figure 4.3** shows a drill bonnet made for accepting a power feed motor. The image also shows slip renewable bushings that could be used for hand drilling

FIGURE 4.3 Drill bonnet.

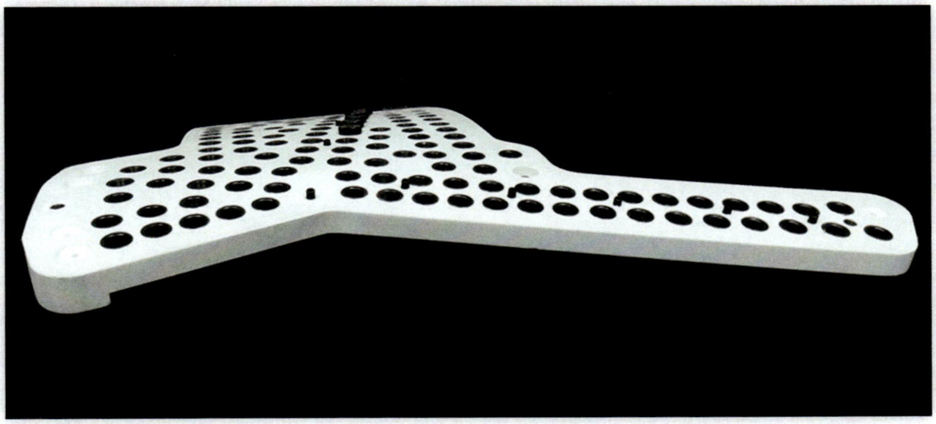

© SAE International/George Bullen

pilot holes prior to application of the power feed motor(s). The drill bonnet is made thicker and more robust because of the strain of holding a large and heavy motor in precise position during the dynamics and strain of the drilling operation. Figure 4.4 shows a power feed motor.

The drill bonnet is made to have space between the surface of the work-piece and its underside. The clearance is incorporated into the construction of the drill bonnet to enable chips to clear away from the work-piece surface. Clearing chips is necessary to prevent the chip from spinning and damaging the surface of the work-piece.

Another feature of the power feed motor is the nosepiece bushing that precisely slips into the press-fit-hardened bushing contained in the drill bonnet, with both having locking features. When the power feed motor is slipped all the way into the drill bonnet bushing, the motor is turned a 1/8th to a ¼th turn to lock the motor into place so it cannot back out of the hole during the drilling operation. Without this feature, the pressure of the drill bit as it progresses through the hole would force the motor out of the bushing. The motor is a stand-alone mechanized motor that has an internal feed mechanism. When hand drilling is performed, the mechanic applies pressure to the back of the drill to push the drill bit against the material to be removed. The internal system of the power feed motor is mechanically driven without forward pressure to neutralize the force of the drill bit against the metal. The interlocking mechanism on the nose of the power feed motor and the drill bonnet mitigate the issue of back pressure.

Advantages of power feed motors include their ability to drill larger size hole-bores in fewer steps and the ability of a single mechanic/technician to operate several power feed motors simultaneously.

The disadvantages of power feed motors include their cost and the increased tooling cost because of their more robust construction and maintenance requirements.

FIGURE 4.4 Power feed motor.

© SAE International/George Bullen

Because of their size and weight, most drill bonnets are attached to the assembly tool. They are precisely located to align with the tooling coordinate system that is in turn aligned with the engineering coordinate system. Drill bonnets are located and held in place by precision L type pins and bolted to the fixture with T pins.

4.4.3 On-assembly Automated Drilling, or Drilling and Countersinking Using a Numeric Controlled Precision Drill and Countersinking End Effector

On-assembly drill and countersinking began when the first Electronic Gantry Applied Drilling Systems (EGADS) went into operation in 1997. **Figure 4.5** shows the first on-assembly EGADS application for drilling and countersinking airframes integrated into the assembly process. These systems were basic systems that were independent actors. They used numeric controlled (NC) commands derived from engineering models to drill and countersink holes. They did not use templates to guide the end-effector tip to the point of intersection with the work-piece. They drilled and countersunk the work-piece in "one-shot," meaning they eliminated the step drill process needed in the hand and mechanized processes. Much of their acceptance and success was derived from integration into the assembly line and a design directive that included close replication of the hand and mechanized processes. They incorporated many unique features that were outside the comfort zone of big machine manufacturers who traditionally use heavy steel work beds and robust gantry type machines to precisely machine a work-piece. Traditional machine processes move work-piece to the machine where it is precisely positioned and aligned with the machine coordinate system and then rigidly held in place by strategically placed camps. The benefit to this approach is that the machine "knows" precisely where the work-piece is located. The disadvantage is the work-piece must be moved to the machine and then back into assembly. Complex parts made from varying materials meant to fit precisely together lose continuity when shuffled around. The best process for these kinds of assemblies is to drill-and-fill them in-place.

The on-assembly drill/countersinking machines are dedicated to a single assembly operation for performing a repetitive task. Each machine is tailored to the application. One challenge was to develop an automated system that could account for the variability of the assembly as it comes together. The solution

© Northrop Grumman

FIGURE 4.5 EGADS, first on-assembly application of drilling and countersinking.

came in the form of a pressure foot that moves forward by means of a precision linear motor. A precision encoder was also attached to the pressure foot. The motor moves the pressure foot slowly toward the surface of the work-piece until the pressure foot contacts the surface and depresses the encoder. The encoder sends a signal to the controller, notifying it that the pressure foot has touched the skin (surface) of the work-piece. The drill countersink operation is begun and depth controlled through the encoders' coordination with the position of the drill and countersink tips. The drill bit and countersink having been preset in direct relationship to the work-piece contact surface of the pressure foot, the engineering model is used to define the NC program that contains the drill/countersinking algorithm and contains the maximum material stack thickness at the coordinates where the drill operation is being performed. Therefore, the pressure foot linear encoder measurement control also defines the exit depth criteria for the cutting tools' removal from the work-piece.

All of these on-assembly drill and countersinking machine attributes smoothed their acceptance, TTP, and proliferation across aerospace. They came in contact, sensed, and touched the surface and provided visible feedback to operators as they performed the formerly hand operations. They were also applied to the assembly operation in which the mechanics, who drilled and countersunk the holes by hand, now operated the machines.

4.5 Application with Replication

The criteria and design directive for application of the new ring laser, noncontact measuring technology provides a measurement solution that closely follows and replicates the current process in each of the process categories: hand, mechanized, and automated drill/countersink. The same technology is incorporated into each solution but packaged with a form factor that enables its transition into each of the currently used process categories with a minimum disruption.

4.5.1 Hand Drill/Countersink Gauging

The design and development of a gauge that can be "freehand" aligned, set, and stabilized into a hole that has been previously drilled and countersunk is challenging. The holes previously drilled are sans template, and all alignment and stabilization to the hole before and during measurement must take place freehand by the mechanic. Therefore, a gauge was developed to facilitate sight, touch, and feel alignment to the hole by means of a spring compressed cone in the front of the gauge that engages the countersink. This attribute is described in detail in Chapter 2, and is similar to the way countersink gauges engage the countersink in the surface of the work-piece today.

The designed product provides a visual lighting system for the operator to clearly see if the gauge is aligned. When aligned, a lighting queue signals for the operator to start the scan. When finished, a lighting queue signals a successful or unsuccessful scan.

If the scan was successful, a go no-go light notifies the operator if the hole was good or bad. All data are collected and stored in a comfortable human-carried case attached to the gauge by means of a cable. The components of the case are battery powered. The operation of the hand system is detailed in Chapter 3.

The form factor of the gauge is ergonomically designed and weighs less than two pounds. The nosepiece positioning, spring type compression, and operation replicated the type of feel and utility of all the former gauges combined into a single gauge. These features facilitate a rapid transition learning curve for the user.

4.5.2 Mechanized Drill/Countersink Gauging

The design and development of the mechanized version of the ring laser noncontact measurement gauge operates the same way as the handheld "freehand" version. The only difference is the nosepiece. The nosepiece has a steel bushing attached that is slipped into the press-fit bushings in the drill bonnet. The nosepiece design incorporates a locking feature exactly like the locking feature on the power feed motors that engage the screws and on the drill bonnet to hold the gauge firmly in place during its operation. This leaves the mechanic free to perform other tasks. A more detailed description is contained in Chapters 2 and 3.

These simple features ensure stability of the gauge during its measurement routine and replicate the process tasks associated with a mechanized drilling operation of the power feed motor. The mechanics learning curve and the comfort level for performing the hole-bore validation process with the new technology are increased.

4.5.3 Automated Drill/Countersink Gauging

Applying a noncontact ring laser measurement gauge and integrating it into an end effector requires specific attention to the design and operation requirements documents provided by both the end user and the integrator. Each application is unique, since the global suppliers of end effectors have arrived at the same conclusions with different solutions, and each application to the specific product requires the assessment and necessary design adjustments (if required) to meet the new requirement. The good news is that all on-assembly drill/countersinking end effectors have to find the surface of the work-piece and incorporate the solution described in Section 4.4.3 of this chapter.

Therefore, if the application requirements document defines the integration into the end effector, positioning the gauge off to the side of the drill motor quill, the same pressure foot solution can be incorporated into the gauge. The precise vector and normal location of the hole just drilled and countersunk is known. Since the centerline of the gauge is mounted and coordinated precisely to the centerline of the drill tip, the motion to bring the gauge into measurement position is performed by a simple offset command in the NC program. All that is left is to perform the "reach-out-and-touch" process previously described for the pressure foot now incorporated into the gauging solution.

The application of this simple solution enables the machine to drill and countersink, move to the gauging position and measure, and if the hole-bore and countersink is a go, move over to the automated fastener installation head and install a fastener. If there is a no-go condition, the machine skips the fastener installation step and proceeds to the next position to begin the process of drill, countersink, measure, and fill.

The application of the single gauge solution empowered by the ring-laser noncontact gauge removes humans from the middle of the drill-and-fill process. The application of the ring-laser noncontact gauge to remove human eyes-on validation of hole-bores and countersinks adds risk that needs to be mitigated by a thorough vetting of each system. The process for selecting, validating, and transitioning a fully automated drill-and-fill solution enabled by a single gauge contains risk that amplifies the need to follow the procedures defined in the book *Automated Drilling and Countersinking of Airframes* [4-2]. The robust validation and TTP process described in that book takes on more importance, since the risk increases for defects beyond just a bad hole-bore or countersink. It includes blinding the hole-bore and countersink from human evaluation (eyes on) by reliance on a single gauge evaluation. In addition, one of the advantages of the ring-laser noncontact measurement system is its ability to provide the hole-bore diagnostics necessary to determine if damage, burrs, or debris were caused from the drilling activity at the mating conjoined surfaces of the stacked materials to be fastened together. In the former process, a second post-drill assessment was made to validate the quality of the hole-bore. In the new process, all criteria for acceptability of the hole-bore and countersink to insert a fastener that will perform as intended rely on a single sweeping evaluation. Validation, testing, and a solid TTP plan need to be in place for such a critical change to the manufacturing process that provides security to the product.

The criticality and singularity of relying on one point of failure to sustain the continuity of the production line must also be considered. Many people, many gauges, and many measurements have been replaced by one gauge and one measurement. The gauge links the drilling and filling operations. Failure of the gauge breaks the chain. Operations (production flow) stop and must have backup manufacturing processes and be rapidly recoverable. The design of the ring-laser noncontact system incorporates in-process calibration, diagnostics, and rapid replacement attributes. One-minute gauge change-outs have been demonstrated to reduce machine downtimes.

4.5.4 Combined Operation—Teaming Application

The use of any expensive piece of automation requires that the machine be used to its fullest capacity so that non-value added tasks can be reduced and the intended functionality of the machine can be realized. The teaming of the automated and handheld gauging process is essential to maximize the application and utility of the automated systems where they are employed. As previously described, the machine skips the fastener installation step and moves to the next hole if the gauge defines the hole-bore or countersink as a no-go. The machine moves on; the hole-bore stays empty pending cause, corrective action, and a fix. Once the rejection cause is determined, repair is performed. The hole-bore and countersink must then be re-evaluated. Stopping a machine and bringing it back for a single operation to re-evaluate the hole-bore and countersink would be a poor use of the asset. Using a handheld gauge to validate the hole-bore and using it as an extension of the automated process enhances the process.

4.6 Application Strategy

The application strategy for proliferation of the new technology was the push/pull approach to test, design evolve, and migrate the gauge into mainstream manufacturing. The approach was global in scope and included primes such as Airbus in Spain and France and Boeing in Everett, WA, USA, and Charleston, SC, USA. They provided the pull as the technology matured and provided the necessary data to increase the confidence in noncontact ring laser gauging capability in the minds of potential users.

The plan included validation at the manufacturing technology labs of the primes while simultaneously providing opportunities for the integrators to provide test samples for the gauge to measure. As the capability of gauge was recognized by the integrators as having potential to enhance their competitive advantage, they began to pull the gauge into their systems' designs.

4.7 Summary

Many times, after presentations at conferences, I have been asked why it is so hard to get a new technology introduced and qualified onto the production line. Companies that provide even simple products, processes, materials, and technologies struggle to get their foot into the door to expose their latest creation to "change the world" of aerospace or automotive manufacturing.

The answers are straightforward and simple.

1. The primes or first-tier suppliers are not looking for your product or do not need it. It is a solution to a problem that does not exist for them. They have identified within their manufacturing processes critical needs. If by chance a supplier hits that mark, it would be rare. Admittedly, aerospace and automotive primes and their tier-one suppliers do a poor job of communicating their needs.
2. You have hit the mark and it is identified as a critical need for the aerospace prime or tier-one supplier to solve a manufacturing process challenge. Good for you. The assumption on the part of many suppliers is that someone in authority within Boeing or Airbus carries around a checkbook waiting for that jewel to appear. Nothing could be further from the truth. There is a clearly defined process tied to a calendar funding rhythm that a supplier needs to understand. The process is a slow growth process that can take years. That is the bad news. The good news is that once your technology, process, material, or product is qualified onto and into the process, it is difficult for someone to dislodge you.
3. Once you (the supplier) have been recognized and funded by the prime or first-tier supplier for a validation of your claims, there is a deep bureaucracy in place to protect the end products to ensure that a secure manufacturing process is not affected by changes. Every effort is made to minimize trauma to the production line. It is incumbent on the supplier to understand in-depth the current manufacturing process to reduce disruption to the current line flow. That is why the noncontact ring-laser measurement Aeroscan™ system is designed the way it is.

The Aeroscan™ ring-laser noncontact laser system was designed to integrate into existing processes and help maximize the utility of other technologies in which it interacts and cooperates. The application of the system is integral and was developed in cooperation with the prime aerospace manufacturers and their integrators to enhance current processes, not interfere with them.

References

4-1. Scott, A., and Hepher, T. "Boeing Reports Wing Cracks on 787 Dreamliners in Production," Rueters Business, March 7, 2014. Available from: http://www.reuters.com/article/us-boeing-cracks-idUSBREA261QG20140308. Accessed January 24, 2016.

4-2. Bullen, G. N. Automated Drilling and Countersinking of Airframes. SAE International. Warrendale, PA, 2013.

4-3. Chillot, R. "The Power of Touch," Psychology Today. 2013. Last reviewed July 16, 2014. Available from: https://www.psychologytoday.com/articles/201302/the-power-touch. Accessed January 24, 2016.

CHAPTER 5

Extensibility

Once a new technology rolls over you, if you're not part of the steamroller, you're part of the road.

—Stewart Brand

5.1 Introduction

Extensibility (not to be confused with forward compatibility) is a systems design principle whose implementation takes future growth into consideration. It is a systemic measure of the ability to extend a system and the level of effort required to implement the extension. Extensions can be through the addition of new functionality or through modification of existing functionality. The central theme is to provide for change—typically enhancements—while minimizing impact to existing system functions.

Extensibility was originally a software concept term that has been expanded to include a broad range of other product components, including the mechanical and electrical elements of an operating system. It incorporates the design principles defined as a system's ability to have new functionality extended, in which the system's internal structure and data flow are minimally or not affected. Particularly, if recompiling or changing the original source code or core technology is unnecessary when changing a system's behavior or adaptation to other applications, either by the creator, other programmers, or systems application engineers. Because software systems and the core technology are long-lived and will be modified for new features and added functionalities demanded by users, extensibility enables developers to expand or add to the software and core technology capabilities and facilitates systematic reuse. Some of its approaches include facilities for allowing users' own program routines or form factor to be inserted or modified to enhance the system's abilities to define new data types and define new formatting markup tags.

It is my belief that a comprehensive understanding of every technology must begin with at least a rudimentary understanding of the historical context for its development.

5.2 From the 60s to Late 70s

In the 60s and through most of the 70s, the United States (US) reigned supreme as the world's supplier of air vehicles. There was no competition and the US possessed the technology, power, influence, and money to sustain the advantage.

Components like thrust reversers were built at Rohr Industries in San Diego, CA, USA, wings at Avco Manufacturing, TN, USA, and 747 center-wing sections at Goodyear Aerospace in Litchfield Park outside Phoenix, AZ, USA. The 747, L1011, and DC10 wide body jets were entering service and everyone wanted one. The free world's military bought American, and the countries behind the Iron Curtain wanted US equipment by purchase or theft of technology.

The parts and components with a few exceptions were built in the US and unified through a common language, measurement (inches), and a common technology for production. Similar to Henry Ford saying, "You could have any color you wanted, as long as it was black," was the air vehicle adage that you could have any airplane you wanted as long as it was American.

The market had become so saturated with US airplanes that manufacturers' departments of countertrade were used to find new buyers by using creative methods to sell to cash-strapped customers. In one case, the Polish government wanted to buy DC10s but had no hard currency; however, they did have provisions (e.g., ham). So Douglas Airplane Company searched the world for buyers of hams so that the Polish could sell hams to get money to buy DC10s.

5.3 Late 70s to Late 80s

The seeds of the current global manufacturing supply chain dilemma, with its complexities and challenges, were sown in the decade between the end of the 70s and the end of the 80s. The seeds were embedded in composites, competition, and countertrade. Composites were emerging as the new material for aerospace and competition was beginning to appear in Europe.

The Airbus A300 began its commercial service entry during 1974. It was followed in the early 1980s by Airbus' shorter-fuselage A310 derivative, then later that decade by the company's single-aisle A320. European military air vehicle manufacturers began to produce US alternatives like the Panavia Tornado that was introduced in 1979. To counter this competitive threat, the US manufacturers' offices of countertrade began to offer manufacturing technology and segments of air vehicles to produce in exchange for buying their air vehicles. In one case, the Swiss bought F/A18 C/D airplanes in part as the result of a technology transfer for coproduction of the F/A 18 C/D vertical stabilizers. Teams of US air vehicle manufacturing experts began to rove the world, teaching airframe manufacturing processes in exchange for buying US products. At first, the parts were small but as foreign expertise grew, countries began to demand larger segments of the airframe to produce. In the US, the market saturation of the 60s and 70s resulted in industry consolidation.

5.4 Late 80s to Late 90s

The next decade saw a dramatic increase in the technology transfer of larger segments of the airframe from prime producers into an ever-widening supply base spreading across the globe. As China and India's economic star rose, offset agreements for technology transfer meant that complex manufacturing plans were not just translated into other languages and measurement systems but also into characters or logograms called hanzi in Chinese or kanzi in Japan. The pieces, parts, and subassemblies produced in the diverse cultural, linguistic, and geographic morass arrived at assembly centers in the US to be sorted out and collectivized into a completed air vehicle.

During this period, further consolidation happened among military and commercial air vehicle manufacturers. This consolidation exposed commercial air vehicle producers to advances in materials developed for military use, specifically composites. They began to transfer the composite materials onto commercial air vehicles. Recognizing the benefits and competitive advantage of the new material, the US manufacturers held the manufacture of composite parts and the processes necessary for their production in-house as core competencies.

5.5 Late 90s to Late 2000s

As composite materials grew as a percentage of air vehicle volume and weight, countries began to play the competition card to acquire composite manufacturing technology. Composite manufacturing facilities opened worldwide, like the TATA Advanced Materials Center for Aerospace in Bangalore, India. Composite manufacturing processes were now transferred from their parent companies in the US to global producers of aerospace parts.

The thickness variability of composite material, and the strict requirements for producibility and stiffness mean that parts arriving in the US and European assembly centers have become increasingly more challenging to manage. Global independent manufacturing databases, data collection methods, measurement systems, language, and systems of writing compound the ability of assembly centers to sort and settle the pieces into a robust and repeatable final assembly.

5.6 Today

Today more than ever, air vehicle manufacturers recognize that the variability of global supply chain manufacturing processes and gauges need a common unifying medium to ensure best fit at final assembly.

The largest component and cost for manufacturing fit-up at assembly are the holes that bind and hold the airframe together. Mitigating variability from a myriad of inspection instruments used for hole/countersink assessment and diagnostics would unify the global supply base. A global supply chain that uses a common solution to identify, inspect, collect, and distribute critical hole/countersink data would facilitate best fit.

During manufacture, data and analytics collected from a common digitally connected gauge would be shared to optimize parts manufactured across the global supply chain. This would ensure that parts fit when joined at assembly.

A noncontact ring-laser gauging solution provides the common gauge and leveraging advanced analytics capability to unite the complex global supply chain used today and mitigates the variability inherent in current diverse manufacturing environments. The Aeroscan™ gauging system applied across the global supply chain would also address the increasingly recognized challenges of drilling/countersinking composites by providing precision and capabilities not present in current gauges.

5.7 Competitive Alternatives

5.7.1 Gauges

The single gauge with enhanced capability specifically designed to capture key measurement and inspection data within the same device is a unique solution to replace the multiple gauge solutions originated for metal applications, and meant to be used by "hand." As highlighted earlier, hole/countersink diagnostic instruments used nowadays offer limited dimensional evaluation and do not lend themselves to simultaneous one-shot drill/countersink and hole/countersink evaluation. In addition, these devices have limited or no ability to collect data for evaluation and predictive analytics.

The variety of handheld instruments used for hole-bore inspection is vast, including specific gauges to measure grip length, diameter, concentricity, circularity, countersink depth, countersink diameter, and perpendicularity. These devices have common limiting features that vary their data results from measurement to measurement and have well-identified shortcomings.

5.7.1.1 Features

- A measuring resolution of 0.0001 in. (0.0025 mm) and accuracy of 0.001 in. that limit reliability by lack of functionally required accuracy
- Direct reading via analog or LCD readout (go/no-go) without data links and analytics software
- Multiple configurations available for a variety of surfaces and hole-bore sizes create complexity and reduce efficiency
- Purported ease of use with minimal training (gauges only) that can vary in the hand of the user
- Statistical process control ready via data port cable is missing

5.7.1.2 Shortcomings

- Measurement repeatability: Studies have shown that there is a 9% spread in repeatability (i.e., different results when measuring the same hole) because of touch sensitivity variations from human to human.

FIGURE 5.1 Metrology devices used for hole-bore inspection.

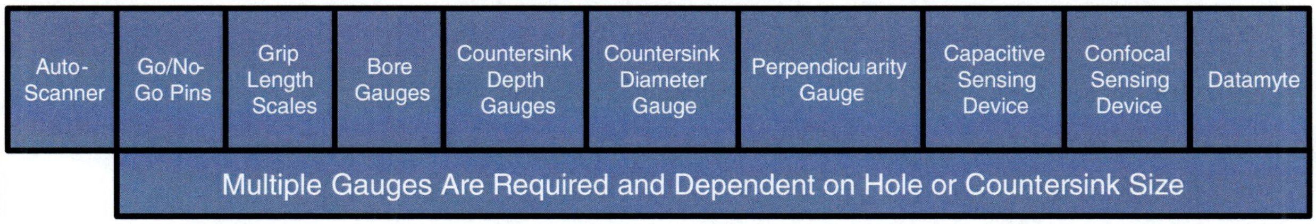

- Require multiple attachments to accommodate a range of hole-bore sizes that increases investment and measurement time (i.e., necessity to change out tips).
- Require direct surface contact to obtain measurement.
- Requires separate device such as a Datamyte™ for data collection.
- Datamyte™ training is costly and time-consuming (e.g., US $1,400/session).
- Requires costly calibration and certification based on predefined Aerospace quality control (QC) standards.

Examples of metrology devices used for hole-bore inspection are described in **Figure 5.1**.

Given the shortcomings of these measurement devices, gauge providers and other metrology instrument producers have continued to invest in alternative technologies. The primary objectives are to address measurement speed, accuracy, repeatability, and comprehensiveness (i.e., perform multiple measurements simultaneously). These alternative technologies include capacitive sensing and confocal sensing.

5.8 Other Emerging Technologies

5.8.1 Capacitive Sensing

Capacitive sensing uses the electrical property of "capacitance" to measure distance. In this tool, the capacitance is measured between the sensor and the airframe surface (e.g., hole-bore wall). The distance between the surfaces is proportional to the measured capacitance. High-performance sensors use small sensing surfaces and, as a result, must be positioned close to the airframe (0.25–2 mm). For hole-bore measurement, a probe tip with capacitive sensors is inserted in a hole to measure the diameter. The measurement is noncontact, fast (~1.5 sec), and highly accurate (+/−0.0003 in.). However, this technology has not been widely adopted in the industry because of a number of inhibitors such as

- Expense (>US $20,000 per device).
- Requires multiple probe tips to accommodate different hole sizes.
- Airframe grounding can cause measurement errors.

- Probe contamination because of the need for close proximity sensing, requiring frequent cleaning and calibration.
- Burrs impact readings.
- They do not capture the full hole-bore profile.

5.8.2 Confocal Sensing

Confocal sensing works by focusing polychromatic white light onto a surface using a multi-lens optical system. The lenses are arranged in such a way that white light is dispersed into monochromatic light. A certain focal distance correlates to a certain wavelength via calibration. Only the wavelength that is exactly focused on the target surface is reflected from the target and passed via a confocal aperture to the receiver, which processes and detects the wavelength changes. These changes are then translated into distance and diameter measurements. This technology has not been widely adopted in the industry for several reasons:

- Expense (>US $60,000 per device and US $120,000 for complete system).
- Requires external ambient lighting.
- The surface to be measured needs to be free from oils or other contaminants such as lubricants commonly used in the drilling operation.
- Scanning composite materials slow down the scanning operation.
- Confocal sensing cannot detect burrs.
- Confocal sensing does not capture full hole-bore profile.

Given the limitations of the current metrology instruments and the lack of adoption of new technologies, the industry has continually sought out new innovations to address the shortcomings of current hole/countersink evaluations. The aerospace industry is vigorously seeking an innovative device that has a broad range of capabilities that address the issues being faced today when performing hole-bore inspection. To illustrate this fact, a comparative analysis of a ring-laser, noncontact single-gauge solution versus other aerospace hole-bore inspection/measurement devices was conducted. Shown in **Figure 5.2**, one holistic solution that accommodates the variety of measurements in a single system and enables data collection/analytics would reduce the complexity of hole-bore evaluation while reducing the number of gauges to perform the analysis.

5.9 Summary of Single Noncontact Gauging Solution

The Aeroscan™ or an Aeroscan™ type device with the same capabilities offers many benefits over the current gauging devices today; notably, the replacement of all current gauging devices for a large range of hole sizes. When considering the costs of the current required gauging devices with all the necessary end pieces for the same range of

FIGURE 5.2 One holistic solution that accommodates the variety of measurements in a single system.

Measurement Devices	Grip Length	Diameter	Concentricity	Circularity	Countersink Depth/ Diameter	Perpendicularity	Hole Quality & Defect Identification	Data Collection	Price
AeroScanner	✓	✓	✓	✓	✓	✓	✓	✓	$30K
Grip Length Gauge	✓								$1.5K - $7K
Diameter Gauge		✓		✓					$1.5K - $7K
Countersink Depth Gauge					✓				$1.5K - $7K
Countersink Diameter Gauge					✓				$1.5K - $7K
Perpendicularity Gauge						✓			$1.5K - $7K
Capacitive Sensing Device	✓	✓							$20K
Confocal Sensing Device		✓			✓		✓	✓	$65K - $130K
Datamyte								✓	$10K – $50K

hole-bore sizes, the total cost of acquisition can easily exceed US $45,000–US $85,000. A single-gauge solution saves time by quickly scanning the hole in a single pass and does not require repeated measurements for holistic hole-bore evaluation.

In summary, the aerospace, road, and rail vehicle has been vigorously seeking a superior hole-bore inspection solution that offers distinct benefits, including truly noncontact, scanning of the entire hole-bore profile in a single pass (instead of discreet points), high accuracy and repeatability, hole-bore quality diagnostics (defects), composite and dissimilar material stack measurement, and advanced analytics at an attractive price.

5.10 Extensibility—Market Size

The extensibility and impact of this type of no-contact ring-laser technology in the aerospace, road, and rail industry will be in the key areas of production, including assembly, subcomponent manufacturing, and tooling. This single-gauge solution will be an integral part of the global supply chain, enabling suppliers and manufacturers to effectively and efficiently manage hole-bore quality. The Aeroscan™ type gauge

- Streamlines the application of hole/countersink evaluation via the integration of several key measurement processes including diameter, grip length, countersink depth, countersink taper, and perpendicularity

- Addresses hole/countersink conformance in an accurate, efficient, and repeatable manner
- Enables automation of the hole inspection process via the noninvasive incorporation of hole inspection as part of the automated drilling process
- Captures data that facilitate advanced and predictive analytics to drive continuous improvement (i.e., predictive drill life)
- Proactively identifies defects such as gapped fasteners via the assessment of perpendicularity prior to fastener placement
- Streamlines the calibration and certification of mechanized and automated drilling systems, drill templates and bonnets, and datum and K-holes

5.11 Addressable Market

To understand the addressable market for a single noncontact hole-bore diagnostic gauge such as Aeroscan™ in the aerospace industry alone, an analysis of the total production and inspection requirements was conducted in each of the industry segments: Aircraft, Helicopters, unmanned aerial vehicles (UAVs), and Rockets and Missiles. In addition, the total population of holes that would be inspected/impacted was identified on a per ship-set (complete aerospace vehicle) basis. The outcome of the analysis was a determination of the total scanners required to service these aerospace industry segments.

The largest sector of the aerospace industry is aircraft production, including the manufacturing of commercial, military, and general aviation (civil) aircraft. The major aircraft manufacturers include Boeing, Airbus Group, Lockheed Martin, Bombardier, Embraer, Northrop Grumman, Dassault Falcon, and Gulfstream. Leading manufacturers in other aerospace segments include Sikorsky, Bell, Augusta Westland, and Raytheon.

The total estimated production across each of these segments for 2017 is

- Aircraft 4,500 ship-sets
- Commercial 1,500 ship-sets
- Military 2,000 ship-sets
- General aviation 1,000 ship-sets
- Helicopters 1,900 ship-sets
- UAVs 15,000 ship-sets
- Rockets and Missiles 7,500 ship-sets

Depending on the size and complexity of the air vehicle, the airframe can contain as many as seven million holes. Holes include predrilled pilot holes where measurements must be taken prior to "up-sizing" the hole to the finished size. To quantify the impact of a single noncontact type gauge on the inspection process, the "addressable market" was defined as the average population of holes that require precision inspection:

- Commercial Aircraft 1,500,000 holes/ship-set
- Military 500,000 holes/ship-set
- General Aviation 500,000 holes/ship-set
- Helicopters 250,000 holes/ship-set
- UAVs 25,000 holes/ship-set
- Rockets and Missiles 25,000 holes/ship-set

The current methods used to inspect, measure, and evaluate holes are labor intensive with varied results. In today's environment, inspectors utilize manual gauges that require contact with the surface. The process and time consumed inspecting the holes vary because of the scope of the inspection, materials, proficiency, and variability of human execution. The process is comprised of

1. Gauge placement.
    a. Separate gauges are required to measure diameter, countersink, grip length, and perpendicularity.
    b. The process must be repeated with each gauge.
2. Insertion of measurement device.
3. 90° counter rotation to ensure completeness of measurement.
4. Reading and collection of data via Datamyte™ or similar device.
5. Flagging of issue as identified (e.g., dimension variation that is beyond specified tolerance). Issue is manually recorded.

The total estimated time for the steps identified above is ≈30 < ≈45 sec. These steps include approximately 15 sec for alignment and approximately 15–30 sec for hole-bore inspection and measurement.

The Aeroscan™ type gauging device is designed to be an integral part of the mechanized/automated drilling processes, thereby eliminating the labor associated with manual inspection. For those holes that are manually drilled and/or are not candidates for automated inspection, the process would be

1. Gauge placement utilizing a guidance system to facilitate fast alignment.
    a. A single Aeroscan™ and a single scan will measure diameter, countersink, grip length, and perpendicularity.
    b. An optical image of the hole will be captured for analysis and defect identification (e.g., burrs, scratches, delamination, fiber tear, and so on).
2. Automatic insertion of measurement device via a trigger function; an actuator controls the speed of insertion (no human effort).
3. 360° full bore measurement: No counter rotation required and no additional human effort.

4. Reading and collection of data integrated in the device.
5. Flagging of issue as identified (e.g., dimension variation that is beyond specified tolerance) and recorded automatically in the data warehouse.

The total estimated time for the steps identified above is ≈13 < ≈18 sec. The steps include 10–15 sec for alignment and 3 sec for hole-bore inspection and measurement.

To assess the total financial impact of the device on the inspection process, there are two primary areas of savings: gauge/measurement tool investment and rationalization of operations/labor.

5.12 Gauge/Measurement Tool Investment

As previously identified, a ring-laser, noncontact measuring gauge such as Aeroscan™ combines several devices into a single comprehensive hole-bore evaluation solution. To achieve the same results, several gauges are now utilized, including diameter, countersink depth/diameter, grip length, and perpendicularity gauges. In addition, a separate device for data collection must be used (e.g., Datamyte™), resulting in a total "solution" cost of ~ US $45,000–US $85,000.

To identify the number of Aeroscan™ type gauges required to service the aerospace industry only, an analysis was conducted to determine the number of gauges and equivalent Aeroscan™ quantities needed to service Boeing's commercial aircraft production hole-bore gauging needs. Tool savings were also projected, comparing total manual tool costs with projected single-gauge solution cost. The results of this analysis, including the extrapolation to the industry segments, are defined as follows.

5.12.1 Manual Gauge Analysis (Current Operating Model)

- Boeing commercial production in 2013: 648 ship-sets
- Total holes inspected per annum (1.5 M × 648): 972,000,000
- Total working days per annum: 200
- Holes inspected daily (972 M/200): 4,860,000
- Time to inspect a hole (manual): 37 sec (avg.)
- Total hours/day and seconds/day (two shifts): 16 h and 57,600 sec
- Expected device utilization: 50%*
- Total daily holes per device (two shifts): 960
- Total devices required (Boeing commercial): 6,244
- Total equivalent devices required/ship-set: ~10 device equivalents**

* includes allowance for productivity, spares, calibration, and certification

** device equivalents include all gauges required to measure diameter, countersink depth/diameter, grip length, and perpendicularity

5.12.2 Single-Gauge Replacement (Aeroscan™ Type) Analysis

- Boeing commercial production in 2013: 648 ship-sets
- Total holes inspected per annum (1.5 M × 648): 972,000,000
- Total working days per annum: 200
- Holes inspected daily (972 M/200): 4,860,000
- Time to inspect a hole (manual): 13–18 sec
- Total hours/day and seconds/day (two shifts): 16 h; 57,600 sec
- Expected device utilization: 50%*
- Total daily holes per device (two shifts): 1,600
- Total devices required (Boeing commercial): 3,037
- Total devices required/ship-set (648): ≈5 devices
- Replacement gauge equivalent holes per annum: ≈300,000

Because of the increased speed, productivity, and breadth of hole-bore inspection and measurement, significantly fewer noncontact (>50% of gauge equivalents) are required.

5.12.3 Industry Analysis

Utilizing the identified Aeroscan™ type gauge device quantities for Boeing, a subsequent analysis was performed to identify the total industry requirement. This analysis assumes manual inspection for comparison purposes. However, the expectation is that the new replacement gauge quantities will be lower because the design of the replacement gauges lends itself to adaptation to automation (i.e., simultaneous inspection with automated and/or mechanized drill/countersink).

5.12.4 Total Projections

- Industry production: 28,900 ship-sets
- Holes requiring precision inspection: 4.8 billion
- Commercial aircraft [(1.5-M-holes) × (1,500 ship-sets)] 2,250 million holes
- Military [(500-K-holes) × (2,000 ship-sets)] 1,000 million holes
- General aviation [(0.5-M-holes) × (1,000 ship-sets)] 500 million holes
- Helicopters [(0.250-K-holes) × (1,900 ship-sets)] 475 million holes
- UAVs [(25-K-holes) × (15,000 ship-sets)] 375 million holes
- Rockets and Missiles [(25-K-holes) × (7,500 ship-sets)] 188 million holes
- Aeroscan™ equivalent gauged holes per annum: ≈300,000
- Total Aeroscan™s required [(4.8 B-holes)/(300 K)]: ≈16,000

The estimated quantity of 16,000 gauges would replace (if a replacement plan were to be implemented) millions of gauges now in use.

As highlighted earlier, one of the distinct advantages of the one-gauge noncontact ring-laser solution is the ability to incorporate inspection into automated drilling processes. Current automated systems drill/countersink and then stop and probe. The automated Aeroscan™ type system, as envisioned, would scan/inspect in-process, significantly decreasing the process time per hole-bore. In addition, the design of the integrated system could allow for simultaneous inspection/drilling on a single device, allowing for even greater productivity improvements. Based on the envisioned design and comparing with the manual process, the expectation is that on a comparative basis with manual gauges the industry would see a 20× productivity increase (i.e., an integrated/automated inspection system could inspect 20 times the numbers of holes of a manual gauge yearly). To understand the impact, projections were made on the total automated Aeroscan™ type gauges required for the industry:

- Holes requiring precision inspection: 4.8 billion
- Percentage of hole-bores candidates for automated drilling/inspection: 20%
- Hole candidates for automated drilling/inspection: 960 million
- Equivalent holes per annum (manual): ≈300,000
- Equivalent holes per annum (automated): 6 million
- Total automated inspection systems: 160
- Remaining manually inspected holes: 3.8 billion
- Total manual gauges: ≈13,000

To define the addressable (extensible) market, target pricing was identified for both manual and automated systems. A number of factors determined market pricing, including comparisons with the existing market and expected financial impact of the device on the industry, specifically the ability to rationalize processes and labor. The pricing and total addressable market are as follows.

5.12.5 Replacing Current Manual Gauges

- Manual gauge current (total costs): US $45K–US $85K
- Aeroscan™ type solution (manual): US $35K
- Total current manual costs/ship-set (10 "devices") US $450K–US $850K
- Total Aeroscan™ type gauge costs (5 devices) US $175K
- Total gauge cost savings/ship-set (Aircraft) US $275K–US $675K
- Total industry savings (16,000 devices) US $160M–US $880M

5.12.6 Incorporating into Automated (Replace Manual)

• Manual solutions:	13,000
• Unit cost:	US $35K
• Total costs:	US $455M
• Aeroscan™ automated solutions:	160
• Cost	US $500K
• Total costs	US $80M
• Total market opportunity (Automated—Manual)	US $535M

5.13 MRO—In Service

Corrosion, fatigue, and impact damage in structure add significantly to the non-routine maintenance burden on air vehicle operators. Non-routine maintenance frequently doubles or even triples the total labor hours expended during a maintenance check.

Adding robust structural design in damage-prone areas, such as passenger and cargo doors in commercial applications, has improved maintenance and repair burdens. Most air vehicles flying today have been designed for bolted repair. The ability to perform bolted repairs in a composite structure is service-proven and offers comparable repair times and skills as employed on metallic airplanes (By design, bolted repairs in a composite structure can be permanent and damage tolerant, just as they can be on a metal structure.). A typical repair in either material may require 24 or more hours of airplane downtime.

All bolted repair begins with removal of the fasteners installed during manufacture to remove the piece of airframe that requires replacement. Repair facilities use hand drill techniques to center the drill in the fastener head. The fastener head is mostly steel, titanium, or other durable and hardened material. The mechanic uses a visual assessment sometimes accompanied by small drill cups to maintain perpendicularity to the airframe surface. The process of hand-fastener removal damages (in many cases) the existing fastener hole and countersink requiring structure, hole, and countersink assessment and analysis to determine the next fastener size to ensure the integrity of the replacement part and the airframe. Undamaged holes and countersinks must also be inspected to validate their conformance with fastener fit for strength and performance criteria.

When the new piece is installed, the first operation is to back drill through the structure holes into the new skin or part. **Figure 5.3** shows a layout of bolted repair. The same repair procedure is used for both composite and metallic parts. The holes are back-drilled to align the existing holes with the new panel holes for best fastener fit. Another inspection is performed to validate hole-bore diameter and all other attributes of the hole to ensure conformance with specified requirements. The panel is then countersunk and the countersinks validated before fastener installation.

FIGURE 5.3 Repair layout of bolted repair of composite structure [5-1].

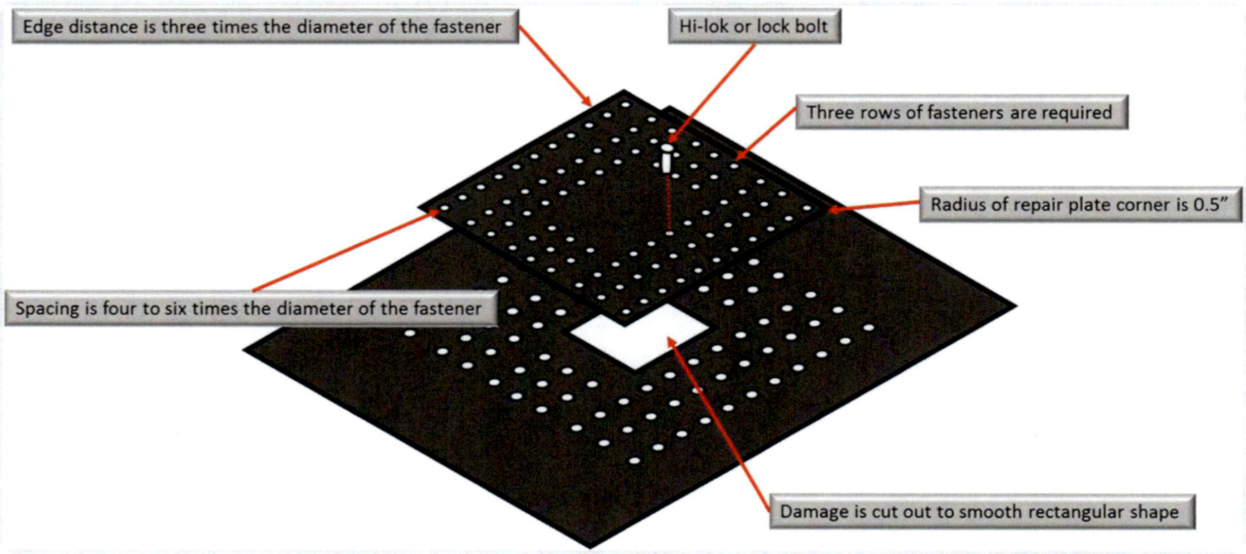

© SAE International/George Bullen

When drilling pilot holes in the composite, the holes for repair fasteners must be a minimum of four diameters from existing fasteners and have a minimum edge distance of three fastener diameters. This is different from the standard practice for aluminum, of allowing a two-diameter distance. Specific pilot hole-bore sizes and drill types to be used should follow specific service repair manual instructions. Composite skins should be backed up to prevent splitting.

Enlarge (back drill) the pilot holes in the patch and parent materials with a drill 1/64 undersize and then ream all holes to the correct size. A tolerance of +0.0025/–0.000 in. is usually recommended for aircraft parts. Figure 5.3 shows a repair layout of bolted repair of composite structure.

The precise nature of repair operations and the critical component for structural strength and integrity require repetitive hole and countersink evaluations and assessment at each step of the process. Multiple gauges are applied, read, and mostly recorded manually to document the repair process as required by law.

Global air vehicle repair (maintenance, repair, and overhaul) facilities operate in environments that range from highly disciplined, environmentally controlled, enclosed facilities to hangers with minimally certified repair mechanics. The range of gauge types and their maintenance used for hole/countersink validation also range from sophisticated to primitive. The variety of gauges needed to fulfill the structural repair of airframes in every case is many. They must be stored, calibrated, and maintained in between use.

A single gauge type like the Aeroscan™ provides a single gauge for optimizing the quality of the repair process and collects data electronically for storage to validate and certify that the repair was done to meet the specified criteria. The added benefit includes electronic data collection for accumulation of a knowledge base for repair procedure improvement, and forensic engineering and evaluation when repairs fail.

The bolted repair level-of-effort primarily resides in the hours to drill-out fasteners and then drill and countersink holes for new fasteners. Up to half of this effort is consumed by evaluation and validation of each hole at each step of the bolted repair process. The application of a single gauge that can perform the evaluation and assessment effort rapidly and then automatically store the data significantly reduces the level of effort.

5.14 Tooling

Aerospace assembly tooling (fixtures) is the controlling medium for precise placement of all parts and components for airframe subassembly through final assembly. Tooling holds parts and components firmly in place by means of a plurality of fiduciary fastening devices. The parts must be held in place by tooling designed to prevent movement during drilling, countersinking, and fastening operations. The function of the assembly tool is the same, whether the drilling, countersinking, and fastening operations are done manually, mechanically, or by automation.

Once fastened together, the subassembly is no longer "green" and can be moved to the next assembly point for mate (joining) to another subassembly, or when completed moved into final processing before shipment to the customer.

Aerospace assembly tooling is highly complex. Its complexity results from the need to consider all aspects of the manufacturing processes, including environmental considerations such as temperature variation to control the coefficient of thermal expansion (CTE) between the tooling and the parts the tooling is meant to hold in place. As the subassemblies are joined and grow in size, tooling becomes larger to accommodate the increasingly larger work-piece.

Controlling part placement with consideration of manufacturing processes and the environment is compounded by floor loading. Large assembly fixtures can weigh in excess of 50 US tons. As a result of the manufacturing processes, environment, and floor loading effects on their as-built precision, assembly fixtures are calibrated semi-annually or annually to ensure that their dimensional fidelity and precision have not been compromised. Calibration holes and targets are placed in and around the assembly tool for tooling QC personnel to use as calibration reference points during the validation process. During this process, the assembly tool is shut down and the surrounding area isolated to maintain a pristine environment for sensitive tooling QC instruments. Calibration can take from a day to several weeks, depending on the fixture size and the degree of work needed to bring the fixture back into conformance.

Attached to the assembly fixture are drill templates (for hand drilling), or drill and countersink test panels for automated drilling and countersinking. In the hand drilling operation, the drill templates control the relationship between the assembly fixture and the collection of parts and components to be drilled. All the holes must be drilled through the stacks of material in their exact position for fastener installation. Controlling elements include the edge distance from the center of hole to the edge of the part(s) to be joined together. The controlling medium to ensure the positional integrity of the drilled holes in relationship to the assembled work-piece are the calibration holes and targets of the assembly tool.

Drill tools are fabricated from softer type materials than the work-piece they drill to prevent scratches to the product during manufacture. The predominant materials are fiberglass or 6061-T6 aluminum. Hardened bushings are pressed into the aluminum or potted into the fiberglass. The hardened bushings are installed to withstand the rigors of the day-to-day operations and repeated cycles of the drilling process. The shop environment and the repeated stress on hardened bushings can degrade the softer surrounding aluminum or fiberglass material or degrade the bushing circumference, causing increasingly out-of-conformance holes.

Drill templates are taken “offline” to tooling maintenance facilities to revalidate and calibrate all their holes, including coordinating holes with the assembly fixture. The drill templates are attached to master models and each hole is tested with go/no-go pins to recertify their conformance with tooling specifications. The master models must be scheduled into the tooling shop and retrieved from storage, setup, and validated prior to their use to validate the drill templates. When the calibration of the drill templates is complete, the master model must be repacked and returned to storage.

In some cases, flat or near-flat drill templates are taken offline and calibrated using coordinate measuring machines that are housed in climate-controlled rooms to maintain dimensional stability and minimize the effects of CTE during the calibration process.

When the drill templates are taken offline, they effectively shut down the manufacturing effort within the assembly cell, unless a costly second set of drill tools is manufactured and is available.

With an Aeroscan™ type gauge available to tooling QC, drill template validation and calibration can be accomplished rapidly “on-station.” In most cases, this would eliminate the need to remove the drill template from its primary purpose of drilling holes, increasing the availability of the critical tooling resource for moving the product through the manufacturing process.

Using this type of gauge for tool calibration on-station, the costly master model removal, unpack, setup, repack, and return to storage operations are significantly reduced.

The added value for improved product quality resides in Aeroscan™'s capability to validate drill templates in place during the production process. Holes can be scanned in the drill templates anytime without isolating or stopping the production process. Hole-bore degradation is identified and corrected through a continuous process rather than discovered by defect or at the time of drill template calibration. The impact to throughput is a direct relationship to the ability of gauge to sustain production operations. Each day the drill template stays online when it formerly was removed from production, is another day (and night) of productivity for the work cell.

An Aeroscan™ type gauge is also immediately available to tooling QC and tooling mechanics for use. The test and operational specifications and requirements for tooling gauges are strict but do not rise to the level of measurement gauges that “touch” the airplane and validate product. The Aeroscan™ gauge can enter the tooling world near term to reduce cost, increase quality, and speed throughput of product.

Automated systems use a test panel(s) attached to and precisely aligned with the assembly tool reference system. The test panels are mounted onto holding fixtures affixed to the assembly fixture. Before any holes are drilled or countersunk into a new subassembly in the work cell, the machine is “zeroed out” using a calibration hole or optical target to ensure the alignment of machine, work-piece, and fixture. Once "zeroed" to the reference system, the machine drills and countersinks several holes into the test panel. The test panel holes are then checked by a series of gauges to ensure their conformance with specifications before being approved to begin drilling on the airframe subassembly.

The Aeroscan™ type gauges add efficiency to this necessary process by making available all the attributes of many gauges, plus adding capability to validate the automated process and protect the airframe.

5.15 Training

Current gauges range from simple mechanical devices to complex instruments that operate as a part of a data collection system. Air, road, and rail vehicle manufacturing plants are highly disciplined environments using trained and certified mechanics to build complex products. The cost of training rises with the complexity of the gauges used to measure and validate holes and countersinks. Training can take mechanics out of the production stream for one to two weeks. The complexity, cost, and training cost of hole/countersink inspection gauges when combined with training and annual recertification downtime mean that they are difficult to migrate for application outside of the disciplined original equipment manufacturers (OEM) and first-tier manufacturers. Within the OEM and the factories of the first-tier air vehicle supplier are specialized, trained, and certified employees whose focus is the inspection and validation of holes and countersinks. They use the gauges daily and become experts in their application and function. The gauges themselves are controlled, maintained, and calibrated to precise standards within these factory environments. While the complexity of current advanced gauges can be controlled within these factories, other environments inhibit and can prohibit their use.

Outside the world of OEM and first-tier suppliers exist lower tier suppliers and MROs, who operate with multifunctional work teams performing a variety of tasks from day to day. The effectiveness and efficiency of these advanced gauges can be defeated by their complexity, requiring weeks of training. When occasionally used, they require detailed work instructions that affect the proficiency of the operators.

A single-use Aeroscan™ is a gauge that adds to the capability of other advanced gauges while providing a simply applied solution requiring a minimal amount of training. These types of single-use gauges do not need regular use to maintain proficiency and incorporate an intuitive operation coupled with straightforward application. The Aeroscan™ type gauge provides a rapid retrieval, use, and replace-to-storage capability that is also ideal for intermittent use outside the large controlled air vehicle manufacturers. It is a solution for MROs and lower tier suppliers that provides the capability beyond current advanced gauges and is as easy to operate as simple mechanical gauges.

Reference

5-1. "Advanced Composite Repair," in *FAA Regulations Handbook*, Chapter 7, 7–45.

CHAPTER 6

Summary

The only important thing in a book is the meaning that it has for you.

—W. Somerset Maugham

6.1 Introduction

United Sciences, LLC ("United Sciences") began its 3-D scanning "journey" by developing an ear plug solution for the United States (US) Department of Defense in response to an improved ear plug to mitigate hearing loss experienced by US warfighters. The hearing loss was the result of suboptimal "one-size-fits-all" type ear plugs made from soft material that was improperly used and did not provide the compact closure to the ear opening required to reduce noise below harmful levels. Innovation and opportunity led engineers to develop and deploy precision 3-D micro-scanning solutions for targeted industries that include air, road, and rail vehicles. Decades of optical scanning and machine vision engineering experience was leveraged to create a noncontact hole and countersink measurement, diagnostics, and analytics solution: The Aeroscan™. Aeroscan™ is a noninvasive disruptive technical advancement in modern air, road, and rail vehicle hole-bore inspection and measurement, and is the only 3-D scanner capable of imaging the interior of a cavity such as the complex hole-bores associated with precision assembly and joining. The Aeroscan™ gauge enables optimized automation in multivariate air, road, and rail vehicle manufacturing, which yields industry cost and time savings, and quality improvement for airplanes, helicopters, unmanned aerial vehicles, missiles, and rockets, as well as road and rail type vehicles.

The critical nature of precision holes that adhere to rigid specifications requires validation and verification. A single fastener failure can cause catastrophic failure in the airframe and can propagate the failure to other fasteners. Since the fastener follows the hole, every element of the hole and countersink is assessed and evaluated for the specific fastener it will accommodate. An airframe can contain as many as seven million holes that must be inspected and verified for conformance to specified tolerances and attributes. As such, the hole-drilling and countersinking operation accounts for as much as 65% of the cost of an airframe and 80% of the defects.

The aerospace industry has led a long and expensive research effort to find a single-gauge solution. The ring-laser optical scanning technology described in this section is the only repeatable, accurate solution that can measure in seconds the depth, dimension, countersink, and perpendicularity of holes varying in size from 3/16 in. (4.8 mm) to 1 in. (25 mm). It is especially well suited for inspection of composite materials, because it is noncontact and provides information that includes fiber tear-out, micro-delamination, and pre-de-stack diagnostics. Aeroscan™ also maintains all captured data for knowledge-based advanced analytics.

A ring-laser type gauge such as Aeroscan™ optimizes quality control (QC) processes and can eliminate unnecessary rework, which will lower cost and improve schedule while eliminating human subjectivity. The core technology provides for implementation in form factors, resulting in easy-to-use handheld solutions or adapted directly on an end effector for use with existing automated and mechanized drilling systems.

The market demand for a single hole-bore gauge solution is expected to be between US $500 million and US $1 billion in annual revenue. The key impact of a single hole-bore gauging solution in fabrication and assembly will be in key areas of production, including airframe assembly, subcomponent manufacturing, and tooling. Aeroscan™ will be an integral part of the aerospace, automotive, and rail vehicle supply chain, enabling suppliers and manufacturers to effectively and efficiently manage hole-bore quality.

Aeroscan™ is the only 3-D scanner that can fit inside a hole and provide noncontact measurement, diagnostics, and analytics of the holes created as a part of the aerospace, road, and rail vehicle manufacturing and assembly process.

6.2 Capabilities

The ring-laser optical scanning gauge has been adapted and modified from core 3-D ear scanning technology. It was developed specifically for the aerospace industry and include

- Optimized optics for aerospace hole-bore sizes
- A laser delivery system optimized to illuminate and image the materials used in the aerospace industry
- Proprietary aerospace hole-bore measurement, diagnostics, and analytics software
- A robust manufacturing and field calibration system
- A hole-bore location system that makes it readily implementable on the manufacturing floor

6.2.1 Measurement Capabilities

The ring-laser optical scanning technology that evolved from the 3-D ear scanner is capable of superior accuracy for measurements, including hole-bore dimension, debris, damage, depth, and perpendicularity (D4+P). For example, hole-bore diameter is measured within 0.0002 in. (0.005 mm). Debris and damage are imaged in real time with accurate identification of the impact of drilling artifacts such as excess liquid shim, resin in the hole, burrs, gaps, cracks, and excess dust. Perpendicularity and grip length are measured with accuracies far exceeding industry standards. Depth or grip length is measured to accuracies within ±0.1 in. Composite-specific measurements are also identifiable such as pre-de-stack analysis, micro-delamination, gaps, and fiber tear-out.

6.2.2 Diagnostics Capabilities

Diagnostic capabilities can provide real-time hole-bore diagnostics, including determining whether a hole is within specified tolerance for D4+P. Repeatability eliminates human subjectivity in evaluating hole-bore diagnostics, enabling consistency across inspectors and inspection locations.

6.2.3 Analytics Capabilities

The system is capable of acquiring, maintaining, and transferring a database of all measurements, diagnostic results, and a complete image of the hole. This data warehousing allows for predictive analytics of tool life, hole-bore forensics, feedback through maintenance and repair operations, and long-term knowledge-based analytics for improved manufacturing tolerances and processes. The system is a closed-loop analytic solution that enables the impact of its use to be quantified and compared for process improvement.

6.2.4 Composite Capabilities

The specific attributes of the core precision 3-D scanning technology is especially well suited for the measurement, diagnostics, and analytics of holes in composite materials. Parts made from composite materials add additional considerations to hole-bore inspection. Parts made from composite materials have thickness variability, are softer, and have dissimilar properties within each drilled hole-bore than the metallic substructure on which they are mounted and attached. Additional issues such as micro-delamination, the need for de-stack, and in-hole fiber tear also complicate the composite part quality assessment process fidelity. Composite materials often have a boundary layer of liquid shim between the composite part and the substructure, which can lead to excess debris between the parts or in the hole-bore after the drilling operation is completed. The noncontact attribute of the gauge avoids contaminants that can degrade efficiency and effectiveness, so it is not affected by debris within the composite hole-bore and optically images the hole, allowing unique composite artifacts to be analyzed.

6.2.5 Speed and Repeatability Capabilities

Each hole-bore measurement and diagnostics is collected in milliseconds (real time), with the measurement, diagnostics, and analytics available for each hole within milliseconds. There is no discernable variability measurement-to-measurement. The repeatability is derived from a machine vision and imaging capability that does not require precision application by the operator. The gauge will repeatedly provide an accurate measurement of the hole-bore as long as the probe is not touching the hole-bore wall.

6.2.6 Multiple Hole-Bore Size Capabilities

The noncontact laser scan gauge technology provides a one-size-fits-all solution. A single gauge is capable of measuring and providing diagnostics for holes ranging in size from 3/16 in. (4.8 mm) to 1 in. (25 mm), without any modifications or changes in parts between holes. This feature provides a gauge to the user that can be used to inspect almost all air, road, and rail vehicle holes.

6.3 Components

The core technology is comprised of the following components.

6.3.1 Optical Probe

The Aeroscan™ gauge uses a proprietary self-actuating optical probe and laser delivery system to illuminate and image the interior of a drilled hole for accurate and repeatable hole-bore measurement and diagnostics. It is the only such probe design and does not exist in any other known 3-D scanner. The laser delivery system is engineered to project a ring of light within the hole-bore for imaging. The laser light is specifically tailored to radially illuminate and image both metal and composite materials, as well as metal–composite stacks of materials.

The probe also includes a wide-angle low-distortion lens structure. The lens structure focuses the direct reflection of the radial light delivered into the hole-bore ring onto a camera inside the scanner. The images captured on the camera are used to image the interior of the hole.

The probe of the gauge is self-actuated and automatically moves into the hole without making contact with either the hole itself or debris within the hole. As such, the probe and the images created through it are not corrupted by contact with the hole or debris. This makes the probe especially useful in measurement and diagnostics of composite holes.

6.3.2 Machine Vision Software Package

The Aeroscan™ gauge includes a machine vision software package for hole-bore imaging and diagnostics. The machine vision software provides real-time D4&P assessment, complete hole-bore imaging, predictive tool life-cycle analysis, and long-term knowledge-based analytics. The software identifies the primary reflection of the ring of illumination light with extreme precision, filters all secondary reflections, and uses the filtered

FIGURE 6.1 Screenshot of a live scan.

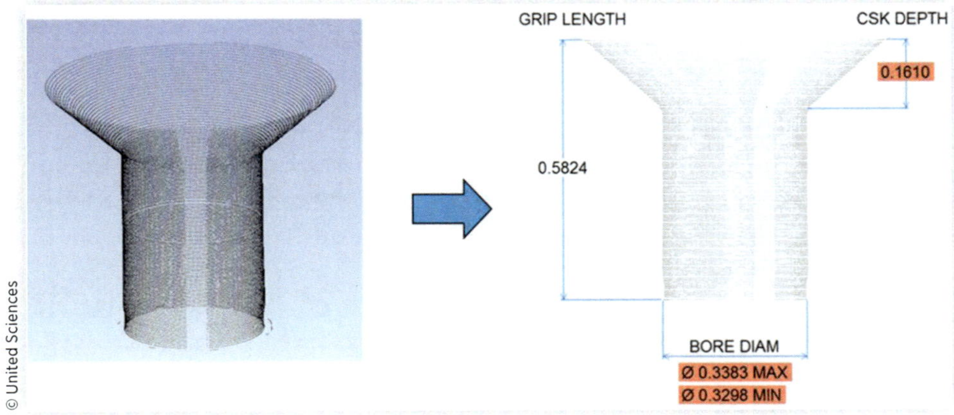

image of the primary reflection to both identify a D4&P assessment and create a 3-D representation of the interior of the hole. From the 3-D representation, the software can provide predictive tool wear analytics previously thought to be cost prohibitive. The software capabilities include identification of micro-delamination in composite materials, burrs, rifling, gaps, and other anomalies in a hole that are now ignored or identified using a cumbersome and costly process. **Figure 6.1** is a screenshot of a live scan, showing the blue illumination reflections and the white software identification of the primary reflection.

6.3.3 Hole-Bore Location Guidance System

The Aeroscan™ gauge is easy to use and requires minimal training for the user. It includes a hole-bore location guidance system that directs the operator to precisely align the gauge over the hole. The hole-bore location guidance system has an alignment camera that provides a live video of the alignment of the gauge and projects cross hairs on the video. An operator visually aligns the cross hairs in the center of the hole to be inspected. This feature has been upgraded to an automated system that replaces the manual alignment. A live video stream is magnified so that the operator can easily confirm the alignment of the gauge over any size hole. When aligned, an operator presses a single button to affect an entire scan of the hole. In an automated application, the gauge is automatically activated through the software when the gauge is centered over the hole. The gauge software automatically actuates the probe, images the hole-bore, and provides a 3-D real-time measurement and diagnostics of the hole-bore. **Figure 6.2** shows

FIGURE 6.2 Screenshot of the location guidance system purposely misaligned to show the hole-bore, alignment cross hairs, and retracted probe tip (far left).

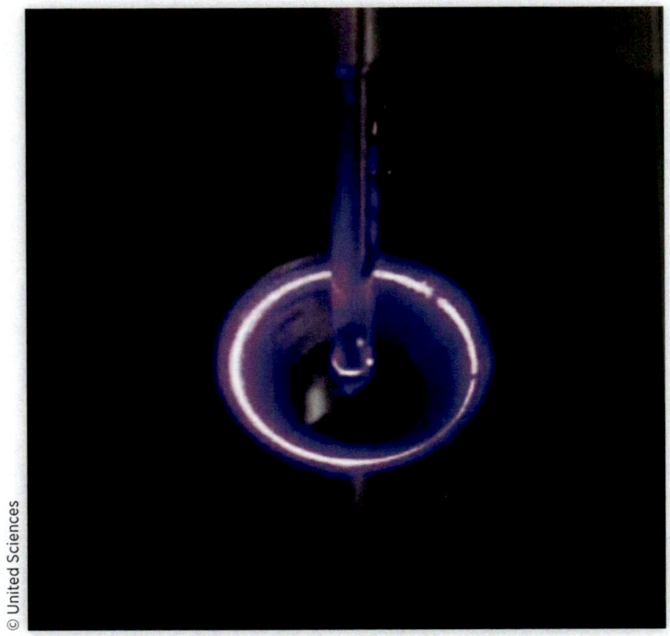

a screenshot of the location guidance system purposely misaligned to show the hole, alignment cross hairs, and retracted probe tip (far left).

6.3.4 Field Calibration Methods

The noncontact laser optical scanning gauge provides quick and long-lasting calibration. The proprietary calibration method establishes a fixed position of the delivered laser illumination with respect to the lens assembly of the probe. The calibration method is quick and may be performed in the field by any operator. Once calibrated, the gauge maintains the calibration for an extended period of time, which is superior to other manual gauges. This enables the gauge to be a repeatable and a continuous source of precise and accurate measurement and diagnostics over a longer meantime between calibrations.

6.3.5 Back-End Solution

The backend solution is designed as an extensible reliable platform, which provides an interface for a variety of data collection, diagnostic, and analytical applications to interface with the aerospace, road, and rail industry ecosystem. The Aeroscan™ system, at its most fundamental level, is designed to manage the workflow of hole-bore inspection and measurement.

A hole/countersink measurement and image acquisition are depicted in **Figure 6.3**, including scanning, the capture of cross-sectional images, and the creation of a 3-D representation of the hole.

FIGURE 6.3 A hole/countersink measurement and image acquisition are depicted, including scanning, the capture of cross-section images, and the creation of a 3-D representation of the hole-bore.

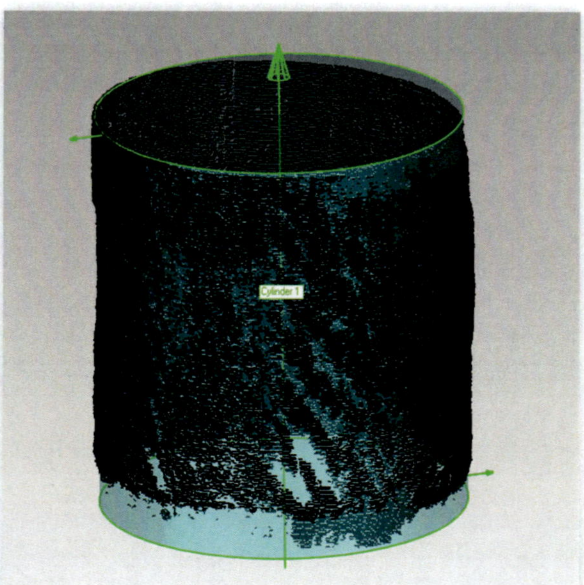

© United Sciences

From an external perspective, it can be seen as a series of web services that provides an application programming interface to allow aerospace manufacturers to interact with a cloud in a safe and secure environment.

Aeroscan™ is not only a gauge, but also a complete end-to-end measurement and diagnostic solution. Such data warehousing of an as-manufactured data of the hole-bore provides short-term QC for manufacturing and long-term hole-bore diagnostic analysis for the improvement of both tolerances and vehicle design. It is a closed-loop analytic solution that allows the actual impact of its use to be quantified.

6.4 Why This Gauge Is a Noninvasive and Disruptive Technology

Air vehicles are the most complex machines made today. Their configuration types are manufactured in small simple shapes and extend to very large sizes with highly complex shapes. Each production line is uniquely tailored with tooling to accommodate the size, configuration, precision, and manufacturing processes to make the rate required to meet demand for the air vehicle type.

The components and parts of air vehicles are made to highly precise dimensional tolerances with little forgiveness for variability, and fabricated from a variety of light-weight high-strength materials. Families of parts are joined together at progressively larger subassemblies until a completed air vehicle is attained. The method for holding all the work-pieces and subassemblies in proper configuration until joined together is assembly tooling.

The primary means used to join all components, parts, and subassemblies of air vehicles are fasteners. Fasteners are inserted into holes that are tightly dimensioned to provide a precise fit for the fastener. The holes may be countersunk or un-countersunk. Countersunk holes are made to insert a flush type fastener and most fasteners are countersunk. The countersunk holes are predominantly on the exposed surface of the airframe, wing, and control surfaces and are used for fasteners to attach skins to the substructure of the air vehicle. They are made flush to facilitate airflow over the surface of the skins they hold tight to the supporting substructure. Depending on the size and complexity of the air vehicle, the airframe can contain as many as seven million holes for fasteners. The hole-drilling and countersinking operation accounts for as much as 65% of the cost of an airframe, 80% of the defects, and 80% of time lost in injuries at the assembly level for an airframe.

Holes are drilled and countersunk in air vehicles using three primary methods. They are automated, mechanized, or hand drilled. The largest percentage of holes drilled today are by hand, followed by mechanized, and then automated. While holes used nowadays are mostly drilled by hand, the increased demand for quality, efficiency, and repeatability has driven more and more aerospace manufacturers to install automated drilling and countersinking on new lines or convert old lines from hand drilling to the automated method of drilling and countersinking holes. Advances in machine controllers, machines/robots, and end effectors have incentivized the move to automation to replace cumbersome tooling and the variability of the hand process.

Fasteners are the primary means by which all structure and skins are held together. Therefore, fasteners are a significant contributor and are critical to the structural strength and integrity of the air vehicle and its ability to satisfy its design purpose. The holes and countersinks for the fit and function of each fastener are the critical component to facilitate optimum performance from each fastener. A single fastener failure can cause catastrophic failure in the assembly it holds together with other fasteners (e.g., the zipper effect). Since the fastener follows the hole, every element of the hole and countersink is assessed and evaluated for its conformance to specification for the specific fastener it will accommodate.

Specified hole/countersink conformance is generally a combination of position and D4&P. D4&P evaluation assesses dimension, depth, damage, debris, and perpendicularity. In aerospace vernacular, perpendicularity is referred to as "normality" or "normal to surface." The number of tools to assess and validate hole/countersink conformance is significant because of the breadth, depth, size, rate requirement, and maturity of the various assembly lines globally producing air vehicles and their components. Specific instruments are used to evaluate each element of the hole-bore. A countersink depth and angle conformance evaluation may use one instrument and a dimensional hole-bore evaluation another instrument, while yet another instrument is used to evaluate the overall depth of the hole-bore (grip length). The air vehicle manufacturing base and the finished air vehicle maintenance, repair, and overhaul (MRO) industry are awash in the variety and complexity of the hole/countersink instruments used today. Some instrument manufacturers have attempted to combine many of the D4&P criteria into a single assessment instrument, but with limited success. The instruments generally have been expensive, complicated to use, and do not incorporate all the D4&P criteria into a single instrument, until the invention of an Aeroscan™ type gauge.

The devices used now have limited or no ability to collect data to populate a knowledge base for analytical assessment and evaluation to improve fit-up at assembly. When applied to automated systems, hole/countersink diagnostic instruments used today offer limited dimensional evaluation and do not provide either simultaneous one-shot hole/countersink evaluation or hole/countersink predictive analytics. The Aeroscan™ gauging system can provide all of these capabilities in a single solution.

The introduction of composite materials onto air vehicles has complicated the traditional hole/countersink assessment criteria due to its finished-part thickness variability, softer and dissimilar properties than the metallic substructure where it is mounted and attached, and the increased attention to other acceptance criteria such as fiber tear in the hole-bore. The addition of composite materials further complicates the assembly process by adding a boundary layer of liquid shim between the composite piece (usually a skin) and the substructure. The liquid shim material is extremely brittle when cured. The use of composites also adds a drill routine that must accommodate drilling through dissimilar stacks of material and "pulling" metallic chips through the softer composite hole. Manufacturers are left with the inefficient manufacturing process of "de-stack" because mechanics cannot assess the interior condition of the composite hole for issues such as fiber tear, damage to the liquid shim, and debris or burrs between the multiple stacks of dissimilar material. In the "de-stack" process, the manufacturer must remove the composite skin after drilling to evaluate the effects of drilling through the various stacks of dissimilar material. This step assures that the parts fit-up without interference from burrs and debris or that the drilling operation has not chipped, cracked, or otherwise damaged the liquid shim.

A device that can satisfy the full range of D4&P criteria and provide a full evaluation of the composite material and liquid shim condition simultaneously, while collecting data to populate a database to facilitate advanced analytics, provides a new disruptive technology to air vehicle manufacturers.

Aeroscan™ is an emerging disruptive hole/countersink technology that provides all the functionality of the multitude of gauges required now for D4&P. It provides the added functionality for composite/liquid shim quality conformance assessment in-the-hole, eliminating the need for de-stack. The single operation provides a nonintrusive and noncontact holistic assessment of the complete hole/countersink, while populating a database for future analytics to enhance the best fit at assembly and variability reduction.

The Aeroscan™ solution can be utilized on assembly in the handheld model or integrated into the automated or mechanized end effector with the automated gauging solution. When incorporated into the automated drilling system, the design operates simultaneously with the drill/countersink operation to maximize utility and output of the valuable automated drilling system. An added functionality in the automated gauging system is its ability to predict hole degradation and tool wear for tool bit replacement to optimize hole-bore quality and drill-bit life.

This functionality of the system can reduce the quantity of drill tools (drill templates and drill bonnets) and their calibration, certification, and maintenance time by inspection of the bushings in the drill tools in place without taking them offline to their master models.

Global MRO and depot repair facilities enhance their ability to service and repair air vehicles that are composed of increasing amounts of composite parts. Repair of composite parts requires a higher level of skill than their metallic counterparts. The holistic diagnostic capability of the Aeroscan™ gauge and its ability to capture data means that information can be sent back to the original equipment manufacturer (OEM) or central repair and maintenance depot for evaluation and best recommendation for repair. Information can be collected in a database to determine recurrent repairs for forensic evaluation and corrective engineering of parts to enhance performance.

In summary, Aeroscan™ is a disruptive technology for hole/countersink evaluation to meet all D4&P assessment, while adding material evaluation criteria and advanced analytics in a single noninvasive short cycle operation. This functionality can replace the array of disparate instruments populating the air vehicle manufacturing and operations environment.

6.5 Strategic Impact

The strategic impact of a noncontact ring-laser 3-D imaging gauge in both handheld and automated configurations is significant and far-reaching. The impact stretches beyond its complete hole/countersink diagnostics, composites evaluation, and analytics capability. It is a complete solution for the air vehicle OEM, their clients, and supply chain to provide one common inspection and measurement solution. The gauge provides a hole/countersink diagnostic instrument that can span the entire air vehicle supply chain to provide commonality, connectivity, communication, and integration. **Figure 6.4** shows the total impact of Aeroscan™ across the total product life cycle for OEM, client, and supply chain.

The functionality is the same whether applied by hand or applied by machine, providing a common instrument and process that are not gauge present in other handheld and automated devices. Current hole/countersink instruments use different devices for handheld and automated hole/countersink assessment, which prohibits integrated data and information transfer across platforms and instruments for factory command,

FIGURE 6.4 Total impact across the total product life cycle for OEM, client, and supply chain.

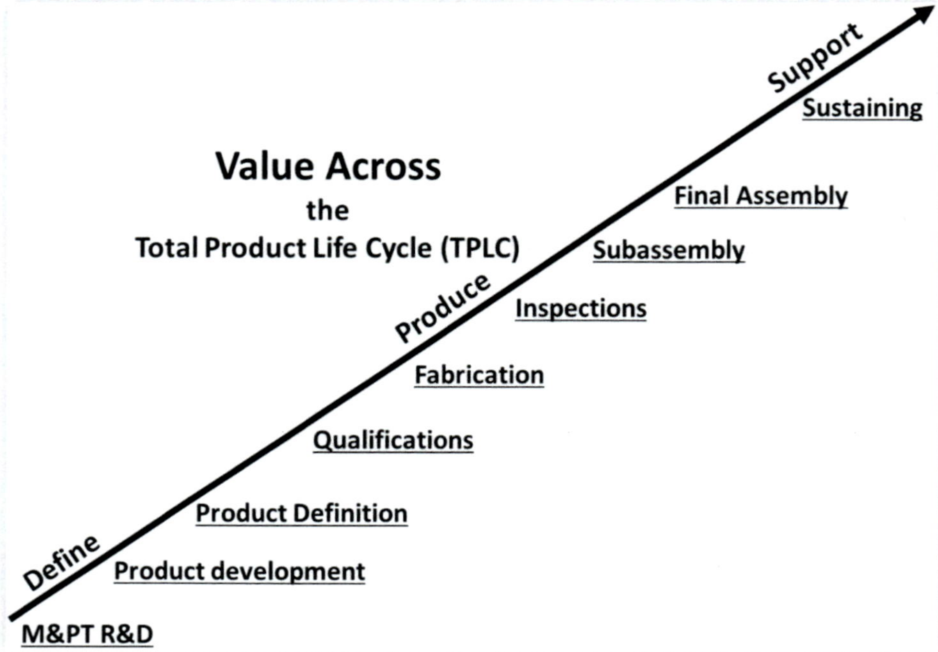

© SAE International/George Bullen

communication, and control. The Aeroscan™ gauge software links automation and handheld devices, providing a complete hole/countersink knowledge-based network across the enterprise value stream.

6.6 Utility

Most handheld gauges for hole/countersink capability continue to grow in complexity, with each additional attribute resulting in training that can cost in excess of a thousand dollars per employee, can last a week, and requires annual retraining. The handheld Aeroscan™'s ease of operation and functionality requires a minimum of familiarity for the operator and, therefore, it can transition rapidly into production.

The Aeroscan™ gauge eliminates the human variability of the hole/countersink diagnostics, measurement, and evaluation process in the mechanics' control by performing the operation for them in one simple application. The handheld gauge simplifies the already taxing job of the mechanic and provides a system that complements their skills to achieve better holes/countersinks. The noncontact application of the gauge is more forgiving and compensates for errors in application that result in misreads, missed reads, and rereads of holes and countersinks. In addition, the noncontact operation of the handheld gauge and its rugged construction makes it robust and prevents damaging scratches resulting from contact with the aero-skin during its use.

Mechanics or the quality personnel have a reduced work burden because the Aeroscan™ solution is applied once to perform simultaneously the actions of several gauges, while simplifying the application action of any one gauge. In addition, the time of measurement and incidence of error are reduced because of the multiple applications of the various gauges now in operation. By streamlining the application of the hole/countersink evaluation, the intrusion time of the QC process is reduced, so other value-added operations are "freed up" to enter the workspace and, therefore, throughput is increased.

6.7 Composite Applications

In its infancy, the performance attributes of composite materials for aerospace parts outweighed the complex manufacturing processes and the material, health, safety, and environmental costs as parts slowly replaced metal. The primary focus was on qualifying composite parts onto airplanes and understanding the complexities of replacing well-understood performance and manufacturing processes of metal materials with parts made from the new composite material. The attributes of composites have and will continue to be the primary driver for their increased use on air vehicles. **Figure 6.5** shows the rapid increase in the use of composites and its continued use as the material of choice on air vehicles.

As more composite parts have made their way into air vehicles, the design, definition, and macro-manufacturing process effects of composite materials have become better understood, but remain challenging. As the material has integrated onto air vehicles,

FIGURE 6.5 Rapid increase in the use of composites and its continued use as the material of choice on air vehicles.

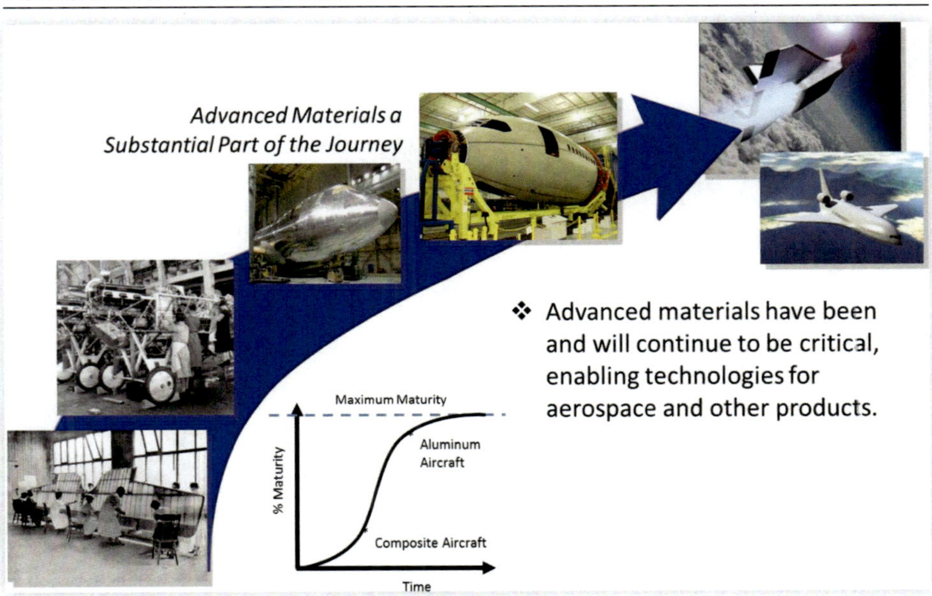

© SAE International/George Bullen

several micro-effects within the manufacturing process have begun to manifest themselves and be recognized as contributors to decreased performance. Damage can occur at several scales within the composite material and structural configuration during the drilling operation. This ranges from damage in the matrix and fiber to broken elements, resulting in failure of bolted attachments. The extent of damage impacts load life and residual strength and, therefore, it is critical to damage tolerance. Fiber breakage can be critical because structures are typically designed to be fiber dominant (i.e., fibers carry most of the loads).

The assessment of in-hole quality has become a critical component during the composite manufacturing process to ensure air vehicle performance and longevity. Aeroscan™'s ability to collect, evaluate, and disseminate critical alerts for tool change or process improvement is unique to it. It stands out as a single solution that can capture and communicate all the necessary measurements for product quality, performance, and longevity. The gauge provides in-process information and data for immediate corrective action based on all the elements of hole-bore quality, including predictive tool changes based on tool wear and in-hole quality.

In an automated application, Aeroscan™ provides a medium for the simultaneous actions of on-assembly hole drilling and evaluation with zero stoppage or interruption. The automated Aeroscan™ gauge operates as a coordinated system to continuously provide data at the last holed drilled without loss of time or motion.

The capability of gauge to evaluate composite in-hole quality and to provide a knowledge base of specific hole-bore quality also facilitates better hand drilling. Specific holes can be identified for manufacturing assessment and refinement to improve hole-bore quality; or mechanics can be identified who may need additional coaching or training for situations where there is no defect, but the hole-bore quality could be enhanced for better product performance.

6.8 Tactical Impact

The improvements in processes and results for hole-bore evaluation using the Aeroscan™ system in air vehicle manufacturing processes result in significant savings in the following areas:

- Hidden costs that are included as a part of the manufacturing process
- Design improvements resulting from reduced manufacturing process variability
- Hand drill/countersink process step reductions resulting from a combination of many tasks formerly requiring many gauges into a single task with one gauge
- Automated process step reductions derived from simultaneous drill/countersink and gauging operations
- Hand and automated hole/countersink touch labor reduction
- Enhanced in-process quality
- Above the shop floor, supports labor reduction
- Cost reductions throughout the supply chain, beginning at the OEM

6.9 Costs of Airframe Manufacturing, Uncollected In-process Costs

There are activities built into and necessary to the drilling, countersinking, and fastener installation process that are hidden within touch labor cost. One example is grip length measurement for fastener selection. Grip length incorporates safety factors for embedment and thread-strip that are important both for the initial installation of the fastener and for long-term reliability with regard to both loosening and fatigue resistance. Grip length assessment for fastener installation was always important but has become more important as composite materials have increasingly made their way into air vehicles. Composite parts complicate grip length determination because their thickness varies by ±5% and in some parts by as much as ±8%.

In metallic assemblies, the mechanic confirms the grip length at every 10 or 20 holes, because the assembled thickness of metallic parts is very consistent. In assemblies that are composed of stacks of material incorporating composites, almost every hole is checked to determine the proper grip length before the proper fastener is selected. The process of sticking a grip length gauge into each hole to determine the proper fastener grip length is built into either the drilling standards or the fastener installation standards. It is a hidden cost and is time and quality dependent, based on the discipline of the mechanic and their visual acuity when reading the gauge. The impact per hole may be only a few seconds when incorporated into the process, but eliminating only two of those seconds per hole when multiplied across one million holes results in 2,000,000 sec, or ~556 h per air vehicle. When producing 600 air vehicles a year, the savings is 333,600 h per year. Using a US $100 per hour rate, the savings is US $33,360,000 per year buried in the process, assuming only a 2-sec reduction in process time. This is an unseen cost that can be reduced or eliminated if grip length is calculated at the time the hole is measured using an Aeroscan™ type gauge.

In the manual grip length assessment, every hole is measured as an independent event based on the composite variability assumption. Data are now not collected to compile patterns in each part geometry to determine areas of repeatable thickness. The mechanic measures the hole, reaches into the appropriate bin in the fastener kit, and installs the fastener. The Aeroscan™ system can collect discrete data and populate a database to perform pattern analysis and map each part for repeatability. This attribute would reduce the variety and number of fasteners stocked at the assembly line and hasten the selection and installation of fasteners. Grip length measurement is only one of the in-process hidden costs that application of Aeroscan™ technology reduces.

6.10 Summary

The ring-laser noncontact optical scanner called Aeroscan™ offers many benefits over the gauging devices used today. Notably, Aeroscan™ can replace all of the current gauging devices for a large range of hole-bore sizes. When considering the costs of the gauging devices, it replaces with all the necessary end pieces for the same range of hole-bore sizes the total cost of ownership for the current gauges, which can easily exceed US $45,000–US $85,000. The Aeroscan™ gauge also saves time by quickly scanning the hole in a single pass, and does not require repeated measurements for holistic hole-bore evaluation.

In sum, the noncontact ring-laser optical scanner is a superior air, road, and rail vehicle hole-bore inspection solution that offers many distinct benefits, including noncontact, scanning of the entire hole-bore profile in a single pass (instead of discreet points), high accuracy and repeatability, hole quality diagnostics (defects), composite and dissimilar material stack measurement, and advanced analytics.

References: Applicable Gauge Patents

6-1. U.S. Patent 8,715,173, filed August 2012; entitled: "Otoscanner with Fan and Ring Laser."

6-2. Allowed for Issuance, U.S. Patent Application 13/767,017, filed February 2013; entitled: "Optical Measurement of Drilled Holes."

6-3. Allowed for Issuance, U.S. Patent Application 14/099,482, filed December 2013; entitled: "Optical Measurement of Drilled Holes."

6-4. PCT Patent Application PCT/US2014/15128, filed February 2014; entitled: "Optical Measurement of Drilled Holes."

6-5. U.S. Patent Application 14/094,130, filed December 2013; entitled: "Optical Scanning and Measurement."

6-6. PCT Patent Application PCT/US14/15676, filed February 2014; entitled: "Optical Scanning and Measurement."

6-7. PCT Patent Application PCT/US14/15674, filed February 2014; entitled: "Optical Measurement of Drilled Holes."

6-8. U.S. Patent Application 14/099,536, filed December 2013; entitled: "Identification of Damaged Tools."

6-9. PCT Patent Application PCT/US14/15670, filed February 2014; entitled: "Identification of Damaged Tools."

6-10. U.S. Patent Application 14/099,567, filed December 2013; entitled: "Profiling a Manufactured Part During its Service Life."

6-11. PCT Patent Application PCT/US14/1565, filed February 2014; entitled: "Profiling a Manufactured Part During its Service Life."

6-12. U.S. Patent Application 13/417,649, filed March 2012; entitled: "Otoscanning with 3D Modeling."

6-13. U.S. Patent Application 13/417,767, filed March 2012; entitled: "Optical Scanning Device."

6-14. U.S. Patent Application 13/586,471, filed August 2012; entitled: "Video Otoscanner with Line-of-Sight Probe and Screen."

6-15. U.S. Patent Application 13/586,459, filed August 2012; entitled: "Otoscanner with Camera for Video and Scanning."

6-16. U.S. Patent Application 13/586,448, filed August 2012; entitled: "Otoscanner with Pressure Sensor for Compliance Measurement."

6-17. U.S. Patent Application 13/586,474, filed August 2012; entitled: "Otoscanner with Safety Warning System."

6-18. U.S. Patent Application 14/040,973 filed September 2013; entitled: "Otoscanner with 3D Imaging and Onboard Tracking."

6-19. U.S. Patent Application 14/041,943 filed September 2013; entitled: "Otoscanner with 3D Imaging and Structure-from-Motion."

6-20. U.S. Patent Application 14/049,542 filed October 2013; entitled: "Tubular Light Guide."

6-21. U.S. Patent Application 14/049,538 filed October 2013; entitled: "Tubular Light Guide."

6-22. U.S. Patent Application 14/049,666 filed October 2013; entitled: "Display for Three-Dimensional Imaging."

6-23. U.S. Patent Application 14/049,530 filed October 2013; entitled: "Fan Light Element."

6-24. U.S. Patent Application 14/049,687 filed October 2013; entitled: "Integrated Tracking with World Modeling."

6-25. U.S. Patent Application 14/049,678 filed October 2013; entitled: "Integrated Tracking with Fiducial-Based Modeling."

6-26. U.S. Patent Application 14/049,504 filed October 2013; entitled: "Integrated Calibration Cradle."

6-27. U.S. Patent Application 14/049,518 filed October 2013; entitled: "Calibration of 3D Scanning Device."

6-28. PCT Patent Application PCT/US2012/029806, filed March 2012; entitled: "Optical Scanning Device."

6-29. PCT Patent Application PCT/US2013/028347, filed February 2013; entitled: "Otoscanning with 3D Modeling."

6-30. PCT Patent Application PCT/US2013/028364, filed February 2013; entitled: "Video Otoscanner with Line-of-Sight Probe and Screen."

6-31. PCT Patent Application PCT/US2013/028323, filed February 2013; entitled: "Otoscanner with Camera for Video and Scanning."

6-32. PCT Patent Application PCT/US2013/028311, filed February 2013; entitled: "Otoscanner with Camera for Video and Scanning."

6-33. PCT Patent Application PCT/US2013/028299, filed February 2013; entitled: "Otoscanner with Pressure Sensor for Compliance Measurement."

6-34. PCT Patent Application PCT/US2013/028218, filed February 2013; entitled: "Otoscanner with Fan and Ring Laser."

6-35. Australian National Phase Patent Application 2012231140 Corresponding to PCT Patent Application PCT/US2012/029806, filed March 2012; entitled: "Optical Scanning Device."

6-36. European National Phase Patent Application 12760319.9 Corresponding to PCT Patent Application PCT/US2012/029806, filed March 2012; entitled: "Optical Scanning Device."

6-37. Chinese National Phase Patent Application Corresponding to PCT Patent Application PCT/US2012/029806, filed March 2012; entitled: "Optical Scanning Device." (Serial No. Not Yet Assigned.)

6-38. South Korean National Phase Patent Application 10-2013-7027859 Corresponding to PCT Patent Application PCT/US2012/029806, filed March 2012; entitled: "Optical Scanning Device."

6-39. Japanese National Phase Patent Application Corresponding to PCT Patent Application PCT/US2012/029806, filed March 2012; entitled: "Optical Scanning Device." (Serial No. Not Yet Assigned.)

6-40. Canadian National Phase Patent Application Corresponding to PCT Patent Application PCT/US2012/029806, filed March 2012; entitled: "Optical Scanning Device." (Serial No. Not Yet Assigned.)

6-41. Brazilian National Phase Patent Application 1120130240652 Corresponding to PCT Patent Application PCT/US2012/029806, filed March 2012; entitled: "Optical Scanning Device."

6-42. Hong Kong National Phase Patent Application Corresponding to PCT Patent Application PCT/US2012/029806, filed March 2012; entitled: "Optical Scanning Device." (Serial No. Not Yet Assigned.)

6-43. U.S. Patent Application 14/109,799, filed December 2013; entitled: "Custom Ear Monitor."

6-44. U.S. Patent Application 14/109,796, filed December 2013; entitled: "In-Ear Wearable Computer."

6-45. U.S. Patent Application 14/109,555, filed December 2013; entitled: "Identity Confirmation Using Wearable Computerized Earpieces and Related Methods."

6-46. U.S. Patent Application 14/109,757, filed December 2013; entitled: "Identity Confirmation Using Wearable Computerized Earpieces and Related Methods."

6-47. U.S. Patent Application 14/109,583, filed December 2013; entitled: "Identity Confirmation Using Wearable Computerized Earpieces and Related Methods."

6-48. U.S. Patent Application 13/958,171, filed August 2013; entitled: "In-Ear Proprioceptive for Relieving TMJ-Related Symptoms."

6-49. U.S. Patent Application 13/958,198, filed August 2013; entitled: "Method of Designing Custom Device for Alleviating TMJ-Related Symptoms."

6-50. U.S. Patent Application 13/958,227, filed August 2013; entitled: "In-Ear Orthotic for Relieving TMJ-Related Symptoms."

6-51. U.S. Patent Application 13/958,258, filed August 2013; entitled: "In-Ear Device with at least One Sensor."

6-52. U.S. Patent Application 13/958,245, filed August 2013; entitled: "Methods for Reducing One or More Symptoms of TMJD (Successive Aural Devices)."

6-53. U.S. Patent Application 13/958,282, filed August 2013; entitled: "Methods for Reducing One or More Symptoms of TMJD (Oral and Aural Combinations)."

6-54. U.S. Patent Application 14/094,210, filed December 2013; entitled: "Administering A Sleep Disorder."

6-55. U.S. Patent Application 14/094,241, filed December 2013; entitled: "Sleep Study."

6-56. U.S. Patent Application 14/094,269, filed December 2013; entitled: "Sleep Disorder Appliance Compliance."

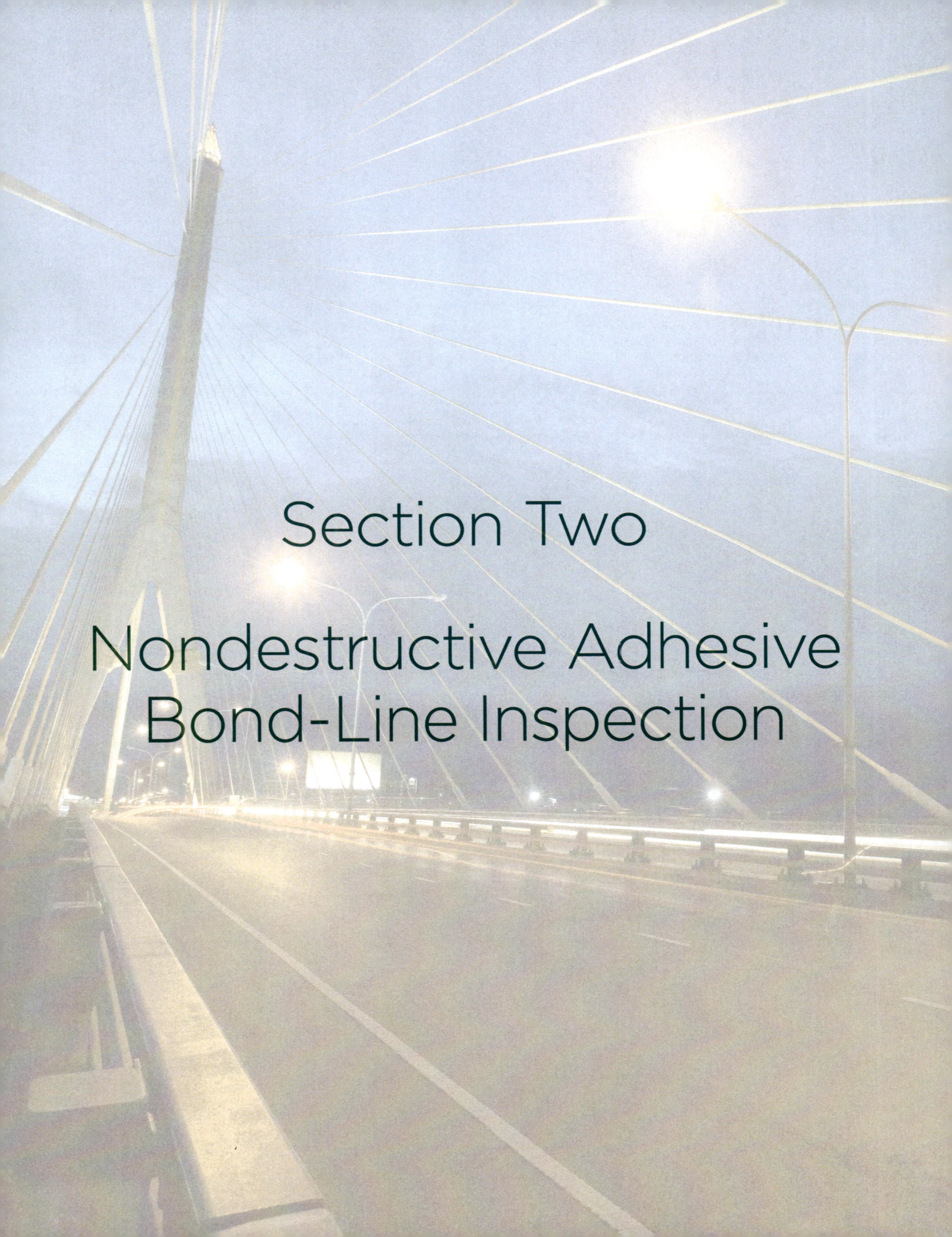

Section Two

Nondestructive Adhesive Bond-Line Inspection

CHAPTER 7

Introduction

Something has to hold it together. I'm saying my prayers to Elmer, the Greek god of glue.

—Tom Robbins

7.1 Introduction

Anyone who has put a plastic model together knows both the ease and challenges of working with glue, and with plastics when the two are combined. Most casual model builders look with pride on their finished construction while wondering how they will get the glue off their fingers, or worse, get their fingers apart.

All the elements are present in these toy models that could be transferable to the large and very real products they represent such as airplanes or automobiles. They have the key holes and pins necessary for determinant[1] assembly. After applying glue to the two mating surfaces of the parts to be assembled, the pins of one part are inserted into the corresponding holes of the other part. When the pins are inserted, they align the seams and all the other control points and aesthetic lines of the mated parts. Rubber bands or other clamping methods are used to hold the two pieces of the glued/mated parts together until the glue dries. A secondary benefit of clamping is to close the gaps created by the part manufacturing method. Every casual and professional model builder knows that the more expensive the model, the better the fit-up. Less warp is present and less surface preparation work is required prior to gluing. Better part manufacturing methods reduce warp, eliminate flashing, and other variations that would require secondary preparation for best fit at

[1] Loosely defined, determinant assembly removes solid fixtures as the primary means of positioning and holding parts together until joined by fastening. Precisely located and strategically placed "Key" holes replace fixtures. In the retail commercial world of furniture, IKEA is a primary and successful user of determinant assembly using the bolt holes of furniture to both position and fasten the parts of an assembly together.

assembly. The same "better" part fabrication processes also produce parts that require fewer post-assembly finishing activities.

This type of determinant assembly "snap-and-glue" construction was not lost on the United States (US) military.

7.2 Models to Machines

The AGM-136A Tacit Rainbow leveraged the basic elements of traditional plastic toy model design and construction to fabricate and assemble the vehicle. AGM-136A Tacit Rainbow was a US military anti-radiation missile program run from 1982 to 1991. Figure 7.1 shows the Northrop AGM-136A Tacit Rainbow in the Cold War Gallery at the National Museum of the US Air Force in Dayton, OH, USA.

The requirement was for a low-cost air-launch system to aid in the destruction of enemy air defense networks. The proposed unit would combine elements of cruise missiles and unmanned aerial vehicles. It would be launched in large numbers by heavy bombers, fighters, or possibly mass ground launch systems. The first test air-launch was on July 30, 1984.

The unit was 8 ft. 4 in. (2.54 m) long with a body diameter of 27 in. (686 mm), significantly larger than the toy models its construction methods would replicate. One process would create molded male V-features along the length of the left and right upper body section. The Vs in the upper body section would match and mate to the molded V groove in the lower body section and be controlled for forward and aft alignment by means of a single pin and hole. Figure 7.2 shows the V tongue and groove construction planned for use in the assembly of the Northrop AGM-136A Tacit Rainbow.

Also molded into the inside upper and lower fuselage sections were grooves that would position and hold the precision numeric controlled machined bulkheads in place and position. Adhesive would be applied to the lower bulkhead positioning grooves prior to their installation of the bulkheads. The bulkheads would be inserted, the unit would be stuffed,[2] and then the upper section would have glue applied its bulkhead grooves; the lower section glue would be applied to the V groove in a simultaneous operation. The upper section would then be flipped and compressed onto the lower section. "Voila!" the parts would come perfectly together. All before the working life of the adhesive expired and invalidated all the work formerly performed. Memories of the old toy model construction days along with its challenges and frustrations emerge from memories long forgotten.

FIGURE 7.1 Northrop AGM-136A Tacit Rainbow.

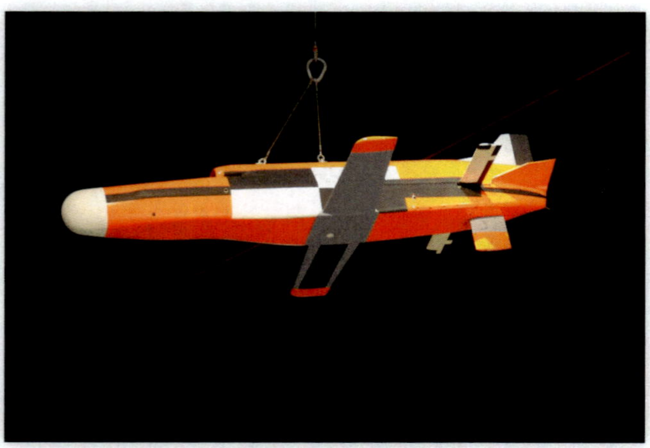

[2] In aerospace assembly jargon, the term "stuffed" refers to the insertion of all the mostly nonstructural components, including cables, tubes, and wires.

FIGURE 7.2 Northrop AGM-136A Tacit Rainbow tongue and groove construction.

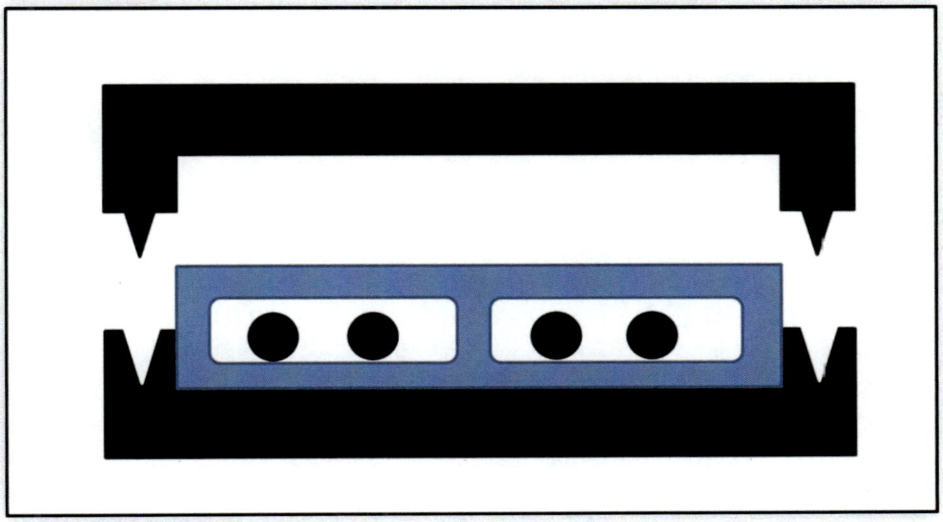

All the processes were intended to be automated using one of the first transfers of automotive assembly automation to the construction of an air vehicle. The process detail elements and working life of the adhesive required an automated process, because a hand method could not be performed fast enough to complete all the assembly operations before the glue (adhesive) lost its viscosity and adhesion properties. The AGM-136A Tacit Rainbow was also intended to be a high-rate production, disposable and terminally guided or launch-and-lost product. Its divergence from the slow-and-low-rate production methods embedded in contemporary airplane pulse lines of the time challenged aerospace production/manufacturing engineers. They were challenged to embrace a moving automated assembly line that replaced the traditional hand assembly methods that dominated the aerospace manufacturing industrial base.

Further challenging the manufacturing engineers was the composite materials incorporated into the AGM-136A construction, and the use of mostly adhesives to replace the tried-and-true fastener to hold the vehicle together.

To summarize, the Northrop AGM-136A Tacit Rainbow was a high-rate production vehicle, using new and unfamiliar composite materials, joined using a method that was time sensitive (adhesives) and contrary to traditional fastening methods, using automation, and positioned together without fixtures using determinant assembly. All these processes were revolutionary (for the time) and attempted simultaneously when none of the processes were familiar or the norm to the people who were to manufacture the vehicle. As it turned out, a bridge too far.

The Naval Research Advisory Committee reported in 1989 that the project was not progressing well. The program was canceled in 1991 (FY 1992), without any production units and at a total cost of around US $4 billion. It was only the second post-Vietnam military project to be canceled after completing testing, but before production.

Many issues contributed to the cancellation of the AGM-136A, including the realization that the toy model construction methods have hidden complexities that need to be addressed before scaling up to a highly complex, precise, and functioning machine such as an air vehicle or an automobile. That is not to say that they are not valid and transferable. Failure is the genesis of success.

7.3 Complexities of Adhesives

In the previous section of this book (**Section One: Noncontact Ring-Laser Metrology Systems**), I identified an emerging ring-laser technology that will enable the complete automation of the drill-and-fill process for fasteners. The technology is indeed revolutionary and applicable for the continued quality improvement, composites use, and increased efficiency for the manufacture of current and near-term air and road vehicles.

The increased use of composites, new aluminum types, development of new adhesives, maturation and efficiency of automated systems, and advanced analytics has driven a fresh look at the toy model manufacturing method for application on complex vehicle types. High cost, complexity, design considerations, and many other challenges of fastening vehicles together are driving vehicle manufacturers to reinvigorate the vision of an all or almost all-bonded structure using determinant assembly.

Global Hawk was another example of an aerospace manufacturer's attempt to transfer adhesives as a primary means of holding a large portion of the vehicle structure together.

The Global Hawk is a high-altitude unmanned reconnaissance air vehicle with high endurance limits. It holds the world record for its class at 33.1 h without refueling. It is used now by the US government in military and humanitarian missions around the world. Figure 7.3 shows an image of the Northrop Grumman Global Hawk, a high-altitude unmanned air vehicle.

FIGURE 7.3 Northrop Grumman Global Hawk is a high-altitude unmanned air vehicle.

© NASA

The Block 10 Global Hawk has a wingspan of 116 ft. (35.4 m) and is 44 ft. long (13.5 m). It has a range as far as 12,000 nautical miles (22,236 km) at altitudes up to 65,000 ft. (19.8 km), flying at speeds approaching 340 knots (about 400 mph) for as long as 35 h. During a typical mission, the aircraft can fly 1,200 mi to an area of interest and remain on station for 24 h.

The Global Hawk aircraft wing is made of advanced high modulus graphite epoxy composite materials and incorporates innovations in design and fabrication of composite aircraft structures. It is the first certified military aircraft that is constructed using an all-bonded primary structure.

"This project is very dependent on the workmanship, on the human factors. No matter how well you control the process, how well you understand every element within the process, you still have that human factor," said Mostafa Pourmand, Northrop Grumman's Structural Integrity lead on the Air Force's Global Hawk, and Navy's Broad Area Maritime Surveillance program [7-1].

The Global Hawk wing ran into one of the common dilemmas of the amateur toy model builder, having patience until the adhesive cures. Every amateur toy model builder wants to see their parts cured and is anxious to move on to the next step in the assembly process as defined in the instructions, if the instruction is used at all. It is common to have a section of the model become a mess when handled before cure, as the builder tries to mate uncured pieces to other uncured pieces of their precious product.

Early in the Global Hawk production process, I was called on to form an action team to address the 7-day takt time[3] for curing the wing skin to the substructure of the Global Hawk wing. The wing was to be undisturbed for 7 full days until the materials process engineers were confident that the adhesion curing process had time enough to complete. The production managers were extremely unhappy with a wing that could not be disturbed and remained idle in a pristine environment until after the 7 days. They and the company were anxious to move on to the next operation that included installation of fasteners. They were impatient and wanted to move on for cost and workflow efficiency considerations.

Many adhesives are highly sensitive to disruption before they are fully cured. Adhesives can lose some or all of their adhesive properties if disturbed prior to the completion of the bonding process. **Figure 7.4** shows the curing process at the molecular level that must take place undisturbed to facilitate a strong bond to meet engineering specifications. Absent the ability (other than destructive tests) to determine the completion of the curing process, the Global Hawk material properties engineers wanted a pristine environment lasting well beyond the normal cure time for absolute assurance that the process was complete. At that time and today, there is not yet a viable technology that can assess the bond-line integrity and cure maturity of a bond-line. This book describes an emerging technology that addresses all the necessary criteria to raise the confidence level of material process engineers to validate a solid bond-line.

[3] Takt time, derived from the German word Taktzeit, translated best as meter, is the average time between the start of production of one unit and the start of production of the next unit, when these production starts are set to match the rate of customer demand.

FIGURE 7.4 Curing of a thermoset adhesive as heat is applied.

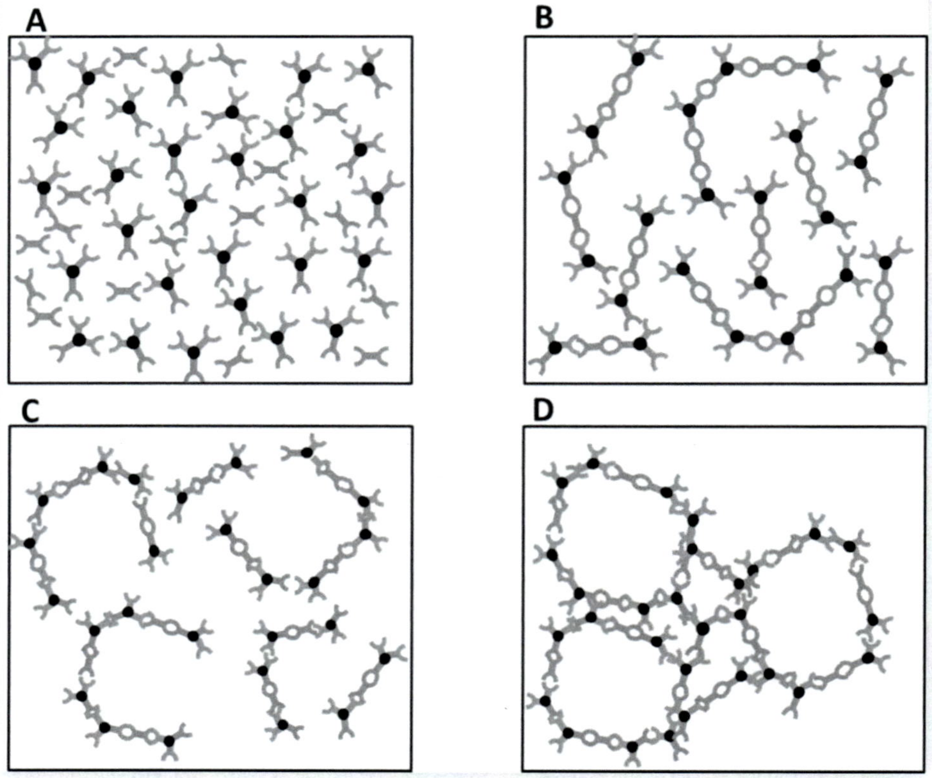

A solid and fully capable bond-line is also dependent on a variety of environmental and manufacturing process activities that need to be tightly controlled to achieve a desired result.

In order for an adhesive to bond (hold together) two surfaces (substrates), there must be several types of interaction between the adhesive and both the substrates. The first type of interaction is that the adhesive must wet the substrate, meaning that the adhesive must spread itself out into a film that covers the substrate surface. In order for this to happen, the adhesive must have a low enough viscosity so that it will flow. Viscosity is the resistance of a liquid to flow. Water has a low viscosity, whereas honey has a high viscosity. Because viscosity is temperature dependent, the application of a cold adhesive to a substrate, or the application of an adhesive to a cold substrate, may result in poor wetting. Another factor that affects wetting is the relative strengths of cohesive forces (between like molecules, such as two adhesive molecules) and those of adhesive forces (between unlike molecules, such as an adhesive molecule and a substrate molecule). If the cohesive forces among adhesive molecules are weaker than the adhesive forces between the adhesive molecules and the substrate surface, then the adhesive molecules will spread out over the substrate and wet its surface. An adhesive that has a relatively low viscosity and is able to wet the substrate surface will flow into any tiny cracks or pores on the substrate surface, thus promoting what is known as mechanical bonding.

Mechanical bonding increases the strength of an adhesive bond and, as a result, a forced separation of the two substrate surfaces is more apt to tear the substrate surfaces. Therefore, surface preparation, including cleanliness, is important to achieving the desired strength properties. Also important is the precise application of the adhesive for maximum contact between the adhesive and the two pieces to be joined.

After extensive work and refinement of the manufacturing process, the action team was able to reduce the takt time for the Global Hawk wing to 2 days.

There is a secondary lesson to be learned that can be derived from the Global Hawk operational experience. It continues to drive home the importance of reducing or eliminating the need for fasteners. I have in all my writings explained the critical nature of fasteners and the potential for a single fastener to cause catastrophic failure of the airplane. It may seem like a simple nuisance, and too costly for you to fix at home. Two US Air Force and Toyota examples of loose screws and improper torque control have recently proven to be costly and critical.

A military headquarters in Afghanistan confirmed that a Northrop Grumman EQ-4B Global Hawk crashed in the southeast city of Jalalabad on August 21, 2011 in a mishap. The EQ-4 is an RQ-4 Global Hawk Block 20 modified to carry the Battlefield Airborne Communication Node. **Figure 7.5** shows the remains of the Northrop Grumman EQ-4B Global Hawk crashed in the southeast city of Jalalabad on August 21, 2011 after it had been reduced to rubble from directed fire to eliminate recovery of sensitive components or systems.

After the Air Force investigated the crash of the Global Hawk aircraft drone that occurred on August 20, 2011, in Afghanistan, improper torque was one contributing

FIGURE 7.5 Image of the remains of the Northrop Grumman EQ-4B Global Hawk crashed in the southeast city of Jalalabad on August 21, 2011.

factor. A loose fastener that was not properly torqued and cost the US Air Force US $73 million [7-2].

"Post-mishap analyses of in-service LRU-X-1 cap screws reflect a significant departure from required values of 18-22 inch-pounds above prevailing torque," reads the report. "Improper torque, insufficient use of thread locking compound or reuse of a deformed lock washer could allow cap screws to vibrate loose during flight operations" [7-3].

It should be noted that the title of the article, "*Loose Wire Caused Afghanistan Global Hawk Crash*" is misleading, since the failure of the fastener that held the wires in place led to the loose wire that caused the crash.

Issues with fastener failure are not limited to air vehicles.

On October 13, 2015, Toyota Motor Sales, USA, Inc. announced that it was conducting a safety recall of approximately 310 Model Year 2016 Tacoma Pickup Trucks [7-4].

The involved vehicles were equipped with a driver's knee air bag module that is attached to the instrument panel reinforcement by four bolts. There is a possibility that incorrect bolts were installed during vehicle assembly at the Toyota Motor Manufacturing de Baja California plant. This could result in the bolts loosening over time. A loosened bolt could affect the performance of the knee air bag, increasing the risk of injury during a crash.

Toyota also announced a recall of certain Tacoma pickup trucks, equipped with an access cab, because of fasteners within the driver and front passenger seat belt assemblies that could loosen if the access doors were repeatedly and forcefully closed over an extended period of time.

If the fasteners loosened completely, the seat belt may not properly restrain the occupant, which would increase the risk of injury in a crash or sudden stop.

Torque control is not something that should be overlooked by manufacturers. It is a critical component in the assembly process; a costly issue if not properly addressed before the assembly process begins. The later a fastening error is detected during the assembly process, the more it costs to correct it. It becomes very costly when failures in a product involve lawyers and lawsuits.

7.4 **Black Metal to Black Gold**

Of course gold is metal. The first part of the title for Section 7.3, *Black Metal to Black Gold* refers to the early replacement of metal parts on a one-to-one basis with composite parts. The second part of the title refers to the evolution of high-strength carbon fiber (HSCF) composite materials to provide much more value than light weight and stiffness. These attributes alone are worth the investment. However, as a deeper understanding of the material has revealed hidden benefits and complexities, the design through manufacturing communities has evolved to recognize, utilize, and address both the benefits and challenges inherent in this complex material.

A higher understanding of the attributes of HSFCFhas led to elimination of fasteners by various methods, including a unified structure and adhesive bonding.

7.4.1 Unified Structure

As understanding of HSFC has matured, design criteria for composite vehicles have evolved to recognize and leverage its unique properties. Fully utilizing composite properties has led to efficient composite structures that are designed differently to spread loads across the structure. The formerly metal stressed members of an all metal automobile substructure have been replaced by composite stressed skin construction shown in **Figure 7.6**. **Figure 7.7** shows BMW's i3 electric vehicle that leverages the advantages of composites to form a carbon fiber-reinforced plastic (CFRP) body.

Designers have taken advantage of HSCF to maximize the composite materials cross sections. They have also learned to use fibers efficiently by varying manufacturing methods such as helical wound parts to achieve torsional stiffness, unidirectional material to achieve longitudinal stiffness, and multidirectional material to achieve multidirectional stiffness.

Experience has led designers of composite structures to increased surface area to spread loads, design for large cross sections within the structure, design for efficient fiber use, and add reinforcement where needed in highly stressed areas to create composite assemblies that complement the properties of the composite itself. They have also recognized the need to always allocate more time for a composite design than a metal design.

A unitized structure reduces mass primarily derived from elimination of longitudinal joints and integration of assembly and separation joints.

A unitized structure also lowers cost by reducing part counts, suppliers, and assembly operations.

While a unitized structure derived from the use of composites has eliminated fasteners and strengthened the structure, there is still a need for a structure to have additional pieces joined or attached to complete the vehicle.

Further fastener reduction could be accomplished through the use of adhesives.

FIGURE 7.6 Differences between a metal designed substructure and a composite stressed skin structure.

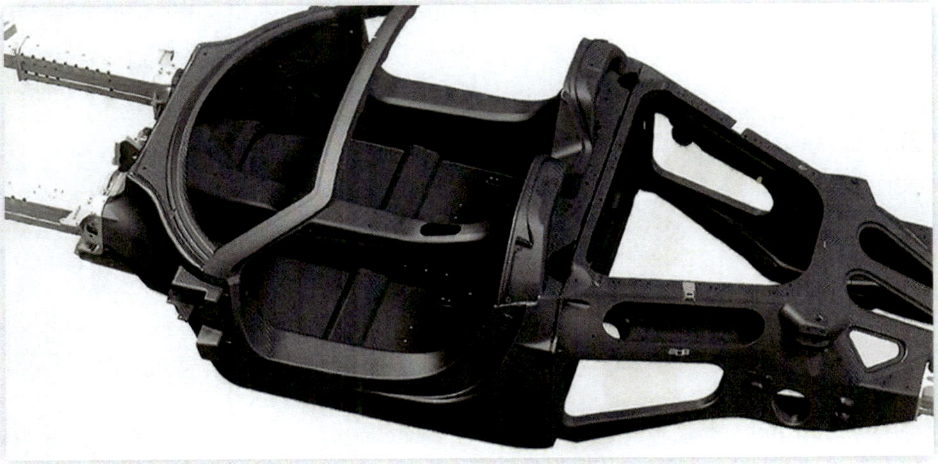

FIGURE 7.7 BMW's i3 uses extensive application of industrialized CFRP.

© BMW

7.4.2 Adhesive Bonding

Adhesives offer a viable alternative to fasteners if a reliable technology was available to provide adequate assurance that the cured bond-line met the specification.

Unitized, adhesively bonded composite primary structures on aircraft could offer substantial cost savings by dramatically reducing part and fastener counts, as well as time and labor for airframe fabrication and assembly. Lockheed Martin Skunk Works (Palmdale, California) demonstrated the possibilities in 2009, during development of the Advanced Composite Cargo Aircraft with a 90% part count reduction in its bonded composite fuselage and vertical tail. However, redundant mechanical fastening is now required for certification of adhesively bonded primary structures on military and commercial aircraft.

Lockheed Martin Aeronautics Co. (LM Aero Palmdale), however, now leads the Transition Reliable Unitized Structure (TRUST) project, part of the Defense Advanced Research Projects Agency's (Washington, DC, USA) Open Manufacturing program. The purpose of TRUST is to develop the manufacturing process control necessary for certification of unitized bonded composite primary structures without redundant fasteners. The goal is bonded primary structure, but no one is confident that bad bonds can be eliminated. TRUST is taking a big data approach, attempting to build confidence through Bayesian process control (BPC). Bayesian logic is a branch of logic applied to decision-making and inferential statistics that deals with probability inference: using the knowledge of prior events to predict future events. There is uncertainty to the application of BPC that leaves future bond-line integrity and viability in the hands of past performance. The future is assumed by a clear understanding of the past. The risk associated with this is that airplanes are complex machines and their manufacture even more complex. Validation is key to assuring that the bonded structure conforms to the required standard.

Another effort to evaluate the benefits of adhesively bonded structure was included under the Rapid Airframe Production Integration Demonstration (RAPID) program, funded by OSD's Defense Wide Manufacturing Science and Technology program and managed by AFRL's Materials and Manufacturing Directorate. Northrop Grumman completed the assembly of a complex, bonded composite, center fuselage structure. The 18-month effort involved conceptual redesign, fabrication, and assembly of a center fuselage section modeled after Northrop Grumman's X47-B. The selected fuselage section is one of the most challenging areas of an aircraft for application of composite structures. The complexity of the bonded structure can be seen in the incorporation of an integrated composite duct and very highly loaded "cat" and "trap" and nose landing gear metallic bulkheads shown in **Figure 7.8**.

RAPID demonstrated methods to fabricate and assemble airframes at lower cost. The RAPID team built extensively on the work performed under AFRL's Composites Affordability Initiative (CAI). Application of out-of-autoclave cure of composite materials, 3-D woven preforms, and bonded joints as well as other individual technologies used throughout the industry has been combined with novel fabrication methods. The end result is a composite airframe that "snaps" together with very little fabrication or assembly tooling compared to conventional composite structures.

Preliminary estimates predict a reduction in nonrecurring tooling costs in excess of 50% and recurring assembly cost reductions of ≈15%. This does not take into account the performance benefits associated with the removal of thousands of airstream fasteners. Fastener count is considerably less with the RAPID configuration versus X-47B configuration. For example, the X-47B configuration upper skin installation consists of 3,879

FIGURE 7.8 Complexity of the bonded structure.

FIGURE 7.9 Paste-bonded skin on the RAPID demonstrator.

© SAE International/George Bullen

fasteners and 7,429 substructure fasteners for a total number of fasteners installed at 11,308. With the RAPID configuration, the total number of fasteners installed is 3,590 resulting in a total fastener count reduction = 11,308 – 3.590 = 7,718.

The concept also enables modifications late in the design cycle that are typically expensive and disruptive. **Figure 7.9** shows the paste-bonded skin on the RAPID demonstrator.

The RAPID project demonstrated the viability of an adhesively bonded structure on complex shapes. RAPID also validated benefits such as reduced cost and lead times. The RAPID program highlighted the need for a technology to rapidly assess and validate the bond-line to move the technologies demonstrated by RAPID into mainstream production process and practices.

7.4.3 Z-Pins

It should be noted that there are other viable and proven alternatives to adhesive bonding of structure. A notable example is Z-pinning to join two pieces of composite structure. Z-pinning structure also helps to validate the cost and weight savings derived from the elimination of fasteners. Mechanical fastener attachment requires pre-curing of multiple details, drilling/countersinking of fastener holes, application of liquid shim, and then the wet installation of fasteners. Attachment with Z-pins requires integration of composite layups, installation of Z-pins prior to cure, and then a backside outer mold line sealing.

The demonstrated benefits are reduced touch labor, reduced weight, part count, defects, increased inter-laminar capability, and improved damage tolerance.

In one application on a small section of the F/A18 E/F, the process resulted in reduced cost of US $85K per ship-set of parts derived from composite part touch labor reduction and weight reduction of 34 lb. per ship-set by elimination of more than 4,800 fasteners per ship-set.

While Z-pinning is not adhesive bonding, it demonstrates the potential saving that could be derived from a viable and verifiable bonded structure to replace fasteners.

7.5 Technology Solution

The development of technologies to address the need to validate a bonded composite structure has steadily progressed, from programs in the late 1990s such as the US Department of Defense (DoD) CAI to more recent initiatives, including the European Union-funded Boltless assembling of Primary Aerospace Composite Structures project. Boeing, Airbus, and LM Aero Palmdale have mounted independent research efforts, offering hope of building reliability into the bonding process, and of gauging final bond strength via a coordinated certification system that includes design, process control, and quality assurance (QA).

Both Boeing and Airbus have identified the need for technology development in several areas that, if realized, could be used to establish a certification process: design, material, and process qualification, process control (termed process safety by Airbus), and nondestructive inspection /nondestructive testing (NDT) methods for QA. There are also issues with durability testing. Durability testing provides assurance that a joint will not fail after repeated loading (fatigue) and under typical in-service conditions (e.g., fluid exposure and temperature and humidity cycling).

Laser bond-line inspection (LBI) uses a pulsed laser beam to create a stress wave that can test the relative strength of the adhesive interface of a composite-to-composite bond. The stress wave strength can be selected by varying the energy and diameter of the laser pulses. Parts are prepared for inspection by applying a thin layer of light-absorbing material such as tape or black paint. Then a thin layer of water is flowed on top. The laser beam passes through the water and is absorbed by the opaque overlay, creating a plasma (ionized gas with no overall electric charge). The water temporarily contains the plasma and intensifies the stress wave, directing it to travel into the material. The compressive stress wave is reflected off the back free surface as a tensile wave, which then travels back to the bond-line, testing its strength (see illustrations on p. 54). This interrogation will leave strong bonds unaffected, but will detect weak and kissing bonds (bond-lines that appear to have intimate contact without mechanical strength).

The technology described in this section was developed by LSP Technologies Inc. (LSPT, Dublin, OH, USA). The technology is a mobile laser bond-line inspection (LBI) system. Inspection is done aided by an articulated robotic arm with a reach of 5.7 ft. (1.7 m). The laser beam used is in the infrared spectrum, at 1,054 nm, with a pulse emitted once in every 8 sec (or 1/8 Hz). The inspection head contains a water delivery and collection system that is held tight to the surface by a vacuum, preventing laser light from escaping. At 5.3 ft. wide by 6.5 ft. high by 13 ft. long ($1.6 \times 2 \times 4$ m^3), the mobile unit fits through standard double doors and is steerable. It can inspect bonded structures up to 1 in. (25.4 mm) in thickness and stores data for later processing. LSPT is developing methods for thicker structures.

The current system was developed to cover a wide range of applications that require pulse energy ranging from 10 to 50 J. The next-generation system will be a more focused portable system applicable to inspect thinner bonded structures, using energies in the 10–20 J range that will reduce the system size.

This type of system and others that are vigorously being pursued will help composites reach their full potential by elimination of fasteners through adhesive bonded-together structure.

7.6 Summary

Composites have flown on commercial aircraft primary structures (those critical to flight) for more than 30 years, but only recently have they been incorporated onto the fuselage, wing box, and wings, most notably on Boeing's 787 Dreamliner and the A350 Airbus (Toulouse, France). These CFRP structures still require assembly with thousands of mechanical fasteners. It is now the easiest and least expensive way to meet certification requirements, which mandate proof that each and every adhesively bonded joint will not separate and cause structural failure should it reach its critical design load. However, many in the industry argue that the full cost and weight savings of composites cannot be realized until bonded joints can be certified without fasteners [7-5].

Adhesive bonding of structures has significant advantages over conventional fastening systems. Bonded joints are considerably more fatigue resistant than mechanically fastened structures [7-6], because of the absence of stress concentrations that occur at fasteners. Joints may be lighter because of the absence of fastener hardware. A major advantage of adhesive bonds, which is often not understood, is that adhesive bonds may be designed and made in such a way that they can be stronger than the ultimate strength of many metals in common use for aircraft construction. In other words, the adhesive will never be the locus of failure for metal-to-metal bonds, provided that the processing is adequate to provide long-term environmental durability. Because of stress concentration effects, mechanically fastened structures will always fracture at the fasteners at a load well below the material ultimate strength.

The disadvantages of an adhesive-bonded joint include the fact that it is a single fastener, and hence, there is no redundancy of load paths unless alternative load paths are added. (Fasteners in an adhesive bond introduce their own problems, which will be discussed in the section on Practical Considerations.) Adhesive bonding is a sensitive process and, therefore, technicians must have appropriate levels of proficiency to perform the bonding process.

A significant limitation of adhesive bonding is that there is no method for providing absolute assurance of bond integrity. All existing quality control tests and NDT inspections satisfy conditions that are necessary for bond integrity, but no test or combination of tests provides assurance of a sufficient condition to guarantee bond integrity. Integrity can only be implied on the basis of quality management and associated QA tests on critical process steps.

A new technology development described in Chapter 9 may be the genesis to provide the evolution from fasteners to adhesively bonded structure as the primary means to hold parts together.

References

7-1. Block, G. Global Hawk Wing Lands on Campus. SDSU News Center. August 24, 2012, December 14, 2009. Available from: http://newscenter.sdsu.edu/sdsu_newscenter/news_story.aspx?sid=73765. Accessed January 18, 2016.

7-2. Mountz, T. Improper Torque & Loose Screws Cause Critical & Costly Issues. In Torque We Share. August 1, 2013. Available from: http://www.mountztorque.com/blog/apply/improper-torque-loose-screws-cause-critical-costly-issues/. Accessed January 18, 2016.

7-3. Rosenberg, Z. Loose Wire Caused Afghanistan Global Hawk Crash. Flightglobal, Aviation Connected: Washington, DC. Available from: https://www.flightglobal.com/news/articles/loose-wire-caused-afghanistan-global-hawk-crash-369238/. Accessed January 21, 2016.

7-4. News Release. Toyota Recalls Certain Tacoma Pickup Trucks. Toyota USA Newsroom. October 13, 2013. Available from: http://corporatenews.pressroom.toyota.com/releases/toyota+tacoma+recall+oct13.htm. Accessed January 21, 2016.

7-5. Gardiner, G. Certification of Bonded Composite Primary Structures. Composites World. March 4, 2013. Available from: http://www.mmsonline.com/articles/certification-of-bonded-composite-primary-structures. Accessed January 21, 2016.

7-6. Hart-Smith, Direct Prediction of Failure in Bonded Composite Lap Joints under Shear Load, 16th International Conference on Aerospace Sciences & Aviation Technology, Military Technical College, Cairo, Egypt. August 1991. Accessed January 21, 2016.

CHAPTER 8

Application

Out of clutter, find simplicity.

—Albert Einstein

8.1 Introduction

Note that in Chapter 7, Introduction to Section Two, the process change from a fastened structure to an adhesively bonded structure is risky and has to include enhanced discipline in the manufacture of the parts to be bonded together, precise and consistently controlled preparation of the surfaces to be joined through adhesive bonding, and a method of nondestructive testing (NDT) that can validate the bond-line without degrading the properties of the adhesive(s) used to hold the two pieces together.

This chapter is about identifying all the elements and requirements necessary for applying a technology that can evaluate the integrity of the bond-line, so that a confidence level is achieved to replace fasteners with adhesives. Assuming the technology described in Chapter 9 is robust and tested, there continues to be concern about whether a chemically bonded structure can sustain its adhesive strength throughout the performance life of the vehicle in which it is used. Adhesives have evolved, but remain suspect over time under the continuous strain, pressure, and extremes of temperature applied during a lifetime of use.

Another concern is fault analysis and failure location identification during final testing. Each successive bond-line is validated as a vehicle is assembled. Using an airplane airframe as a case example, the fuel tanks are pressurized to validate the strength of the bonded assembly. One "sign" that a joint has failed is derived from a very audible

bang sound. When the airplane is assembled, it would be very difficult to identify the location of the failure. Once depressurized, the pieces previously adhesively bonded close up, and failure point(s) are difficult to locate and repair. And the size and structure of the adhesive bond-line inspection technology limits its access to the interior of the fuel tank to re-evaluate the bond-line. This failure condition created through pressure testing has driven a vigorous search for technologies that complement bond-line inspection technology to advance airframe and vehicle assembly toward fastener reduction or elimination.

However, the journey to eliminate fasteners using an adhesively bonded structure begins with validating the adhesive bond-line.

The first step for applying a new technology solution is to identify the requirements necessary to validate the ability of the technology to mitigate all the concerns raised by the process change.

8.2 Composite Bonding Definitions

Before beginning the requirements definition for applying a technology to validate two adhesively bonded structures, a short explanation of what is meant by "adhesively bonded structure" is necessary. There has been some confusion about the definition of an adhesively bonded structure, which can be clarified by defining the various methods for joining composite parts together.

8.2.1 Co-curing

Co-curing is a process of curing a composite laminate and simultaneously bonding it to some other uncured material, or to a core material such as balsa, honeycomb, or foam core. With this process, all resins and adhesives are cured during the same process.

8.2.2 Co-bonding

Co-bonding is a process in which two or more parts (elements) are joined together where at least one of the parts is fully cured and at least one is uncured. This process requires careful surface preparation of the previously cured substrate with additional adhesive(s) required at the interface.

8.2.3 Adhesive Bonding (Secondary Bonding)

Adhesive bonding, also called secondary bonding, is the process of joining together, by the process of adhesive bonding, two or more pre-cured composite parts, where the only chemical or thermal reaction occurring is the curing of the adhesive itself.

Adhesive bonding requires very careful preparation of each previously cured substrate at the surfaces to be joined together. These are called the bonded surfaces. The adhesive bonding process requires well-designed fixtures to align and clamp the parts together during the adhesive bonding process. While it is a temptation to increase the rate of cure of the adhesive to improve cycle time, re-heating previously cured substrates can be risky and has had catastrophic results where parts have been scrapped.

The proper amount of adhesive applied for complete squeeze-out for total coverage of the areas to be joined together is essential. Adhesive application must be done without too much over squeeze-out, and when done by hand is a developmental process derived from learning and repetition. When applying adhesive to surfaces that must be joined together without the benefit of fasteners, automation may be the only solution that provides the adhesive dispensing uniformity necessary to ensure a robust bond-line.

The automotive industry uses a combination of robots with adhesive dispensers attached that are guided by laser or optics to control the adhesive distribution. Bead inspection technology measures and inspects the bead height, width (volume), and position to avoid leaks, voids, and ensure structural integrity once the parts are adhesively bonded together. Dispenser "monitors" also alert users where material is deposited with insufficient volumes by detecting skips, including partial skips.

8.3 Metal Bonding

Metal bonding is the same as secondary bonding (composites), except that with metal bonding the substrates are metal instead of cured composite substrates.

Occasionally, metals are bonded directly to composites using one or more processes. Metals require very stringent surface preparation, including application of corrosion-inhibiting primer prior to bonding to obtain long-term bond durability at the metallic interface. As a caution, care must be taken when bonding metal to carbon as galvanic corrosion can occur in the metal substrate.

A significant difference exists between bonded joints in metallic and composite structures. It is possible to design a joint between metals so that the adhesive is always stronger than the metal being joined. This is not always possible with composites because of the significantly higher strength of composite materials. Also, the first-ply peel failure mode of composites, where the relatively low inter-laminar shear or peel strength of the resin system causes failure of the composite before failure of the bond occurs.

8.4 Adhesive Bond Function Theory

It is necessary to understand how adhesive bonds function to understand the causes of adhesive bond failures, and the importance and sequence of preparation processes prior to adhesive bonding.

The four basic theories of adhesive bonding are as follows [8-1]:

1. The mechanical interlock theory, which attributes adhesive bond strength to surface roughness
2. The diffusion model, which attributes adhesive bond strength to the diffusion of molecules at the interface of the adhesive at the bonding surface
3. The electrostatic attractive model, which attributes adhesive bond strength to weak molecular attraction
4. The adsorption theory, which attributes adhesive bond strength to a combination of weak molecular attraction and chemical bonds (ionic, covalent, and metallic) between the adhesive and the surface of the adherends

While surface roughness has been shown to be a factor in adhesion [8-2], the mechanism of adhesion is not just a mechanical interlock. The adsorption theory is most widely accepted as representing the actual behavior of adhesive bonds. Chemical reactions associated with the adsorption theory can be used to provide a mechanistic explanation for short-term bond strength while also explaining the gradual degradation of these bonds by hydration of the interface [8-3]. According to the adsorption theory of adhesion, adhesive bonds depend on the formation of chemical bonds between the adhesive and the surface of the adherends at the interface. These chemical bonds may be ionic, covalent, metallic, or electrostatically attractive. This theory explains why adhesive bonds on poorly prepared surfaces initially demonstrate high bond strength but have poor service durability. The surface preparation process used during manufacture determines the strength of the interface.

Adhesion is related to chemical reactions, therefore, the surfaces to be bonded must be free of contaminants that would hinder the formation of chemical bonds. The surfaces must also be chemically active to enable the reactions to occur. Since solvent wiping alone does not change the chemical reactivity of most surfaces, solvent cleaning alone is not sufficient to enable development of an adhesive bond. Additional processes are required to produce a chemically active surface. The processes include abrasion or chemical etching to remove the chemically stable surface from the adherends.

Even if a surface is clean and chemically active, these conditions alone are not sufficient to ensure long-term bond integrity. Many bonds formed on surfaces thought to be clean and chemically active have failed in later service, usually because of hydration of an oxide layer as the interface is attacked by water. A classic example of this mechanism occurs on the surface of bonds to aluminum. If the surface is solvent degreased and abraded prior to bonding, adequate strength is demonstrated if the joint is tested soon after manufacture. If, however, the joint is allowed to stand, then the strength of the joint will degrade with time. The explanation for this phenomenon is that the chemical bonds formed during adhesive cure are disrupted as the aluminum oxide layer hydrates. As the oxide hydrates, the interface weakens, leading to a loss of bond strength. If completed, these chemical reactions will eventually result in a total loss of bond strength. Such surfaces exhibit fully interfacial (adhesion) failure with all adhesive left on one surface and none on the other. An essential feature of any surface preparation process is that it must produce a surface that is resistant to hydration.

8.5 Basic Bonding Requirements

Three conditions are necessary for proper surface preparation for adhesive bonding:

1. The surface must be free of contamination.
2. The surface must be chemically active.
3. The surface must be resistant to hydration.

These conditions also influence the sequence of processing. A surface must be degreased prior to chemical or abrasion treatment to produce a chemically active surface. If the surface is degreased after exposing the active surface, the cleaning process will spread any contamination over the entire surface. It has been shown [8-2] that solvent degreasing of a surface will result in a monolayer of solvent residue that can inhibit chemical reaction. Similarly, if the surface is not cleaned prior to treatment to produce a chemically active surface, contamination will either prevent intimate contact with the etchant or will be spread by abrasion. Since the chemical activity of a surface degrades with time, the need to minimize surface exposure time after treatment is essential.

8.6 Bond Strength and Durability

A primary contributing factor to poorly performing adhesive bonds is the failure of manufacturers to differentiate between bond strength and bond durability. Some bonding shops used to rely on the lap-shear strength test, ASTM D 1002, as a quality assurance test for bonding processes. ASTM D 1002 [8-4] has been used to test for selection of surface preparation for the bonding processes.

Strength and fatigue tests may show that the bond strength is adequate or the structure is sound and has sufficient fatigue resistance at the time of testing, but they do not verify that the component will be durable throughout its service life. Bond durability is dependent on the resistance of the adhesive-to-adherend interface to attack by water. Water causes hydration of the surface layer atoms and, in the process, breaks the chemical bonds between the adhesive and the substrate, thus leading to an interfacial failure of the bond. The surface preparation process performed at the time the components are bonded establishes the resistance to hydration.

Three possible outcomes after adhesive bonding, depending on the integrity of the process and the integrity of the implementation of the process, are as follows:

1. A badly prepared or contaminated surface will not be chemically active, will have poor short-term strength, and will exhibit exceptionally poor durability. These problems can be identified by lap-shear tests performed as a quality control measure.

2. A poorly prepared surface may be chemically active and form a high number of chemical bonds, which give good short-term strength, but if the surface is not resistant to hydration, the bond strength will gradually deteriorate to zero. This is typical of a poorly validated process that is performed well. This standard of bond usually passes any short-term quality control tests and even fatigue tests.
3. A well-prepared surface will be chemically active, and a high number of chemical bonds are formed with the adhesive. This type of bond will be resistant to hydration. The bonds will maintain high strength for a long period of service and will also pass quality control and fatigue tests.

The difficulty in acceptance of bonded structures is the inability to differentiate between a poorly selected surface treatment that has been prepared well and a surface prepared with a validated reliable surface preparation that has also been performed well. Static strength and fatigue testing have formerly been used to demonstrate airworthiness [8-5]. These tests do not interrogate the environmental durability of an adhesive bond. It is possible that bonded structures that meet certification requirements may fail in service because of degradation caused by unreliable surface preparation processes used during manufacturing. Because surface preparation is critical to the durability of an adhesive bond and degraded adhesive bonds may exhibit negligible strength, surface preparation must be considered as a critical element in any certification program for flight structures.

The fact that bond degradation is unrelated to loads has been demonstrated by failures of bonded sandwich panels. Spare bonded panels removed from storage have failed even before the components were fitted to an aircraft or road vehicle. The assumption that adhesive bonds fail because of loads is incorrect. This assumption propagates the belief that structures that experience low loads do not need extensive surface preparation. Many examples exist in which this assumption has led to failure of the bond. Adhesive fillers have and continue to be added to panels to dampen vibrations. Because filler is not expected to carry loads, only a simple surface preparation is used during bond preparation. The subsequent dis-bond of the insert material leads to water entrapment and severe corrosion of the underlying metal. Demonstration of the resistance of the interface to environmental degradation (hydration) produced by the surface preparation processes is critical to the safety of bonded structures.

8.7 Understanding Bond Failures

Development and substantiation of adhesive-bonded structures requires a basic understanding of failure modes for an adhesive bond. Adhesive-bonded aircraft structures have been in service for a long time because of their light weight and superior fatigue performance over mechanically fastened structures. However, bonded structures and joints have demonstrated considerable variation in the reliability of service performance. Dis-bonds in joints and sandwich panels have occasionally been incorrectly attributed to a wide range of causes, including fatigue [8-6]. The type of adhesive was often changed in an attempt to improve the performance of bonded components. The end result was frequently bond failure.

Failure to recognize the real causes of bond failures has enabled the continued use of deficient bonding processes, in the continued manufacture of defective components and the use of poor repair technology [8-7]. The misunderstanding of bond failure mechanisms has meant that inappropriate test methods continue to be used for selection of bonding materials and processes. It is important to establish robust and repeatable bonding practices by identifying the real causes of adhesive bond failure. Clear identification of the failure mode plays an important role in determining the causes of bond failure. Correct and reliable identification of the causes of adhesive bond failures is critical to the safety of bonded structures.

8.8 Identification of Adhesive Bond Failures

An adhesive-bonded joint may fail in three basic ways:

1. Within one of the adherends that is outside the joint (this is a desirable outcome because the adhesive joint has achieved the maximum possible performance)
2. Through the adhesive layer (cohesion[1] failure) when the strength of the adhesive bond is exceeded
3. At the interface between the adhesive and one of the adherends (adhesion failure), usually at a load well below the design strength of the bond

Failure of the adherends outside the joint is achievable [8-8] for well-designed and fabricated adhesive bonds in moderately thin adherend materials. This condition is highly desirable because it may reduce the testing required for structural certification [8-9]. If the adhesive has a load capacity that is higher than the strength of the surrounding structure, then only the surrounding structure needs to be tested to demonstrate structural integrity. These type of bonds are also damage tolerant and can be designed using simple design procedures [8-9].

Cohesion bond failures result in fracture of the adhesive and are characterized by the clear presence of adhesive material on the matching faces of both adherends. Failure usually occurs by shear, but peel stresses or a combination of shear and peel may also cause a cohesion failure. This type of failure typically occurs in lap-shear tests such as ASTM D 1002 or peel tests such as ASTM D 1781-76. In cohesion failures, the adhesive surface typically appears rough and may have a lighter color than the bulk adhesive material.

[1] The common practice is to designate interfacial failure of a bond in the adjectival form "adhesive" to describe interfacial failure. This risks confusion with failure of the adhesive material, which is termed "cohesive" failure. In order to more clearly distinguish the forms of failure, the author advocates the use of the words "adhesion" and "cohesion."

FIGURE 8.1 Types of adhesive bond failure.

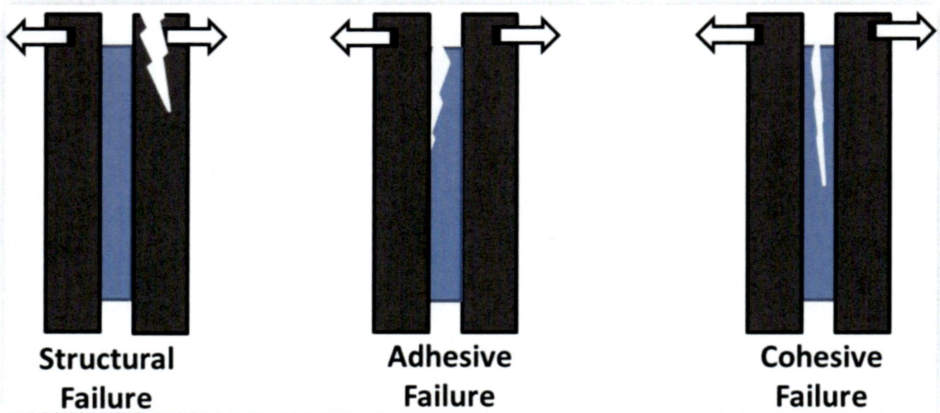

In film adhesive systems, failure usually occurs along the plane of the carrier cloth (a material added during the production of the adhesive roll to aid handling during use). Cohesion failures found in service are typically caused by poor joint design (insufficient overlap length or excessive peel), although excessive porosity will also result in cohesion failure. Figure 8.1 shows the three types of failure, including cohesive failure.

Adhesive bonds are very fatigue resistant [8-10], and only under certain circumstances will a cohesion failure be caused by fatigue. The Primary Adhesively Bonded Structure Technology program [8-11] showed that fatigue of the bond should never occur on thin materials, such as those typically used for aircraft skins, provided the joint has sufficient overlap length. The assumption that adhesive bonds usually fail by fatigue has been perpetuated by old Federal Aviation Administration (FAA) training publications [8-6].

Fatigue failures in adhesive bonds usually only occur where the structure being joined is thick and loads are sufficiently high. Fatigue failures always occur through the adhesive in a cohesion failure. Fatigue failures typically do not occur at the interface; however, interfacial fatigue failures that do occur are symptomatic of a degraded interface. For film adhesives, fatigue failures usually propagate through the plane of the carrier cloth.

8.9 Adhesion Failures

Adhesion failures can be caused in three ways:

1. Ineffective generation of a chemically active surface because of contamination of the bonding surfaces or ineffective performance of a surface preparation process during production
2. Inability of the surface preparation to produce a chemically active surface that is resistant to hydration because of inappropriate process selection and validation
3. Adhesive curing before the bond was formed

Manufacturing process problems or selection of poor processes leads to in-service adhesion failures. Other causes (peel stresses because of operating loads, fatigue, and adhesive creep) may contribute to the final separation of two components that would have eventually failed by some other means.

Contamination, aged adhesive, and poor processing can usually be detected by the low strength of companion coupon lap-shear tests accomplished during fabrication, and better fabricators have quality systems in place to prevent these failures. The same production quality sequence will approve an inappropriate process, provided that it was performed well. These are the failures usually detected in-service after the bond has degraded. If more reliable validation tests were used in the process selection, the processes that produce inadequate long-term durability would be eliminated. A nondestructive inspection method to detect deficient bonds that will subsequently fail during service does not yet exist. **Figure 8.2** shows examples of adhesion failure.

Care is required to correctly assess surfaces that show mixed adhesion and cohesion failure. Interfacial degradation occurs over time, and if a partially degraded bond is subjected to high load, the weakened interface may fail, overloading the adhesive in the regions that have not fully degraded. This will appear to be a mixed mode failure (see **Figure 8.3**) in which the adhesion failure regions exhibit an absence of adhesive while the cohesion failure regions exhibit some adhesive fracture. If the surface shows some adhesion failure and the adhesive has fractured close to the surface (and not in the carrier cloth) on the cohesion failure regions, then the failure is an adhesion failure that occurred before the interface could fully degrade. Cohesion failures always occur in the plane of the adhesive carrier cloth.

FIGURE 8.2 Types of adhesion failure.

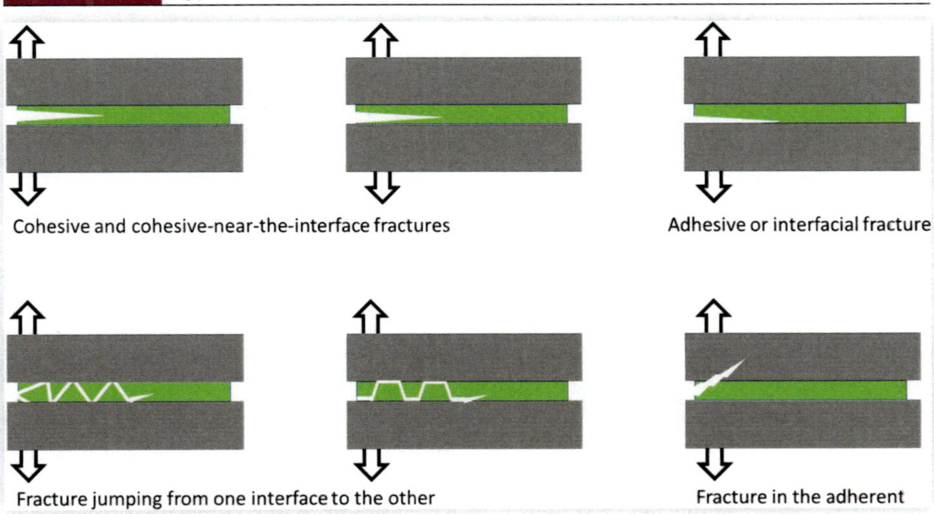

FIGURE 8.3 Mixed failure mode.

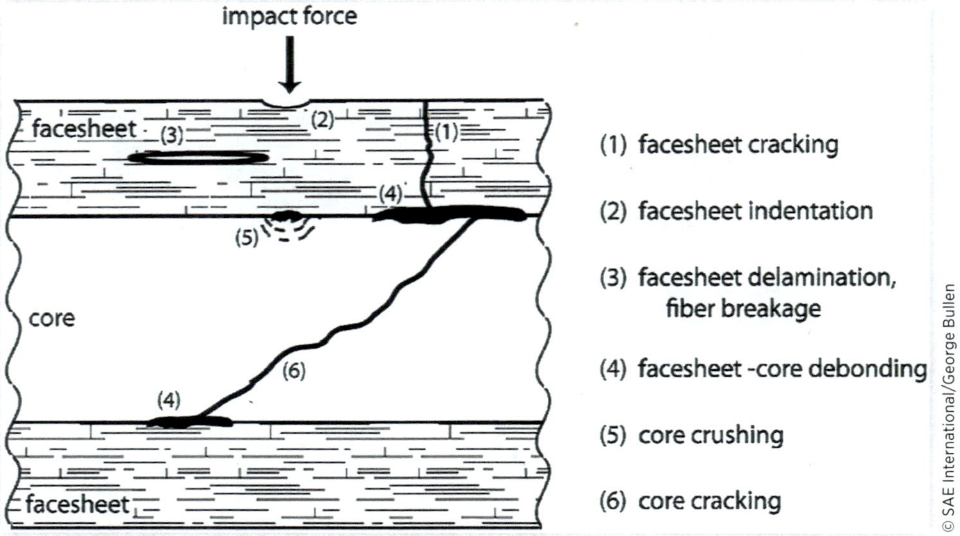

8.10 Failure of Composite Bonds

Adhesive bonds to composite materials may fail in the same manner as adhesive bonds to metallic structures—by either adhesion or cohesion failures. The causes of these failures are often similar to those already stated for failures of bonds to metallic structures. An additional cause of adhesion failures on composite surfaces is the use of inappropriate peel plies.

Extreme caution is required if peel plies are used to protect composite bonding surfaces from contamination prior to bonding. Many commonly available peel plies are known to use fibers that have been coated with release agents so that the peel ply can be successfully removed from the surface of the composite without damaging the top ply. Typically, these release agents are silicone or Tetrafluoroethylene based. These release agents are transferred onto the bonding surface from the peel ply [8-12]—violating one of the basic requirements for a durable bond—that the surface is free of contamination. There is high risk that the interface between the adhesive and the composite will be weakened and susceptible to moisture ingress. A light grit blast will significantly reduce the residue of release material from the surface [8-12].

A second method for prevention of adhesion of the peel-ply fibers is the use of corona discharge-treated fibers. That treatment causes the fibers to be glazed, leaving a slick surface that will not bond to the resin system. This system does not transfer release products to the surface, but leaves a cast impression of the glazed surface on the composite, violating another of the basic requirements for a durable adhesive bond that the glazed surface is not chemically active. The glazed surface is chemically inactive and difficult to bond [8-13]. A reliable method for improvement of the bond interface is to lightly grit blast the surface after removal of the peel ply.

In some cases, the use of a peel ply as the only surface preparation of a composite part prior to bonding has generated durable bonds. The chemical nature of the adhesive and composite surfaces causes a strong acid—base reaction, overcoming the chemical inertness of the composite surface. Careful selection of the peel plies must be made to assure that these chemical reactions do occur and are repeatable, thus requiring a reliable test of bond durability. Demonstration of the long-term durability of surfaces prepared by use of peel plies is critical to the safety of bonded structures.

Another form of failure in bonded composite joints occurs when the top layer of the laminate is broken away from the remainder of the laminate. This may occur when the applied shear loads exceed the inter-laminar shear strength or where excessive peel stresses are encountered. Demonstration by test or analysis that surface peel-ply failure of bonds to composite structures is critical to the safety of bonded structures.

8.11 Summary: Substantiation of Structural Integrity

Methods for substantiating a bonded-joint design are usually based on analysis and/or testing to assess compliance with a design standard. The usual method for certifying composite and adhesive-bonded structures is to rely on a building block approach, whereby strength is demonstrated at coupon level followed by supplemental testing, followed by full-scale structural testing. These tests are in addition to those for the development of design data. The number and complexity of the tests are intended to develop an acceptable level of confidence in structural integrity of the parts to be adhesively joined.

Because of the variability of adhesive properties, a number of tests are undertaken at the coupon stage with fewer supplemental tests and a limited number of full-scale tests. Many tests are essential for any design based on the average shear stress methodology and for any design in which the adhesive is the locus of failure.

The number of tests can be substantially reduced if the design can demonstrate that the structural integrity is independent of the adhesive bond. By ensuring that the adhesive load capacity exceeds the design load by an appropriate margin of safety and providing adequate overlap length to enable the adhesive to achieve that load capacity, the adhesive will rarely fail. If met, then all failures will occur in the structure away from the joint. Using a reliable design methodology may significantly reduce the cost of substantiation of a design. The test program only need establish, with the proposed joint configuration where the bond does not fail before the structure does. This can be achieved with sufficient tests to establish that the load capacity and overlap length are sufficient to achieve the design strength to a statistically acceptable level. Because the variability of structural materials is substantially lower than the variability of adhesive bonds, an acceptable level of confidence can be achieved with considerably fewer specimens.

Understanding that interfacial degradation is a primary failure mode for adhesive bonds on inadequately prepared surfaces, an important element of any substantiation test program must be failure-mode identification. Because a joint achieves nominal design strength after manufacture does not mean that the structure will sustain that strength during service life. The opposite is true in many cases. Structures that show adequate strength during testing have failed in service. Assessment of the failure mode is fundamental to understanding why a structure fails. The focus of testing is placed on loads at failure. However, the failure mode will often foretell how a structure will perform in service. A fundamental rule rigorously applied is that any interfacial failure occurring during certification testing must be considered a failure to certify the structure. The failure is not structural if the surface preparation procedures were inadequate and will always result in service failures when the interface hydrates. A separate test program prior to production, regardless of substantiation testing, should validate the durability of the surface preparation. Elimination of the causes of every interfacial failure during certification testing is critical to the safety of bonded structures.

Any substantiation program for a bonded structure must address the effect of typical service damage on structural integrity. Wherever possible, typical damage types should be identified during the certification program, so that structural integrity can be demonstrated in the presence of real or artificial defects to substantiate the analytical models of damage. Misunderstood in substantiation testing is the demonstration of damage tolerance of adhesive bonds. Many tests are undertaken using Teflon inserts to demonstrate that the structure can sustain bond line defects of a specified size. This approach is only valid for production defects such as macro-voids, because if the surface preparation next to the Teflon insert is effective and if the design is adequate, then the defect should not propagate.

With dis-bond growth in adhesive bonds, the propagation of cohesion dis-bonds is totally different from the growth of adhesion dis-bonds. Cohesion dis-bonds are driven by stresses and are usually the result of poor design (thermal stresses and inadequate overlap length) or possibly production bond line voids. Adhesion dis-bonds are a direct result of failure of the interface because of either contamination during production or degradation by environmental effects. Fracture mechanics can predict the growth of cohesion dis-bonds, but this type of analysis is inappropriate for prediction of the behavior of adhesion failure. Therefore, representation of dis-bonds by Teflon inserts must be restricted only to the demonstration of structural integrity postproduction. If a dis-bond is found in service that was not identified by postproduction NDT and no obvious damage initiated the defect, then it must be assumed that the dis-bond is interfacial and a sign of degradation that cannot fixed by local repair. Also, it cannot be simulated during substantiation testing by Teflon inserts, because propagation from Teflon inserts will be a cohesion-type failure, not adhesion.

Recognition that in-service dis-bond growth is a cause for removal of the repair is critical to the safety of bonded structures.

The following chapter describes a technology that is new and evolving as a method for validating the adhesive bond-line of adhesively bonded structure. It is not without considerable opposition and concern that the aerospace industry is looking to adhesively bonded structure as a replacement for fasteners as the primary means to hold air and land

FIGURE 8.4 The animated reconstruction of the in-flight failure.

© SAE International/George Bullen

vehicles together. Without a nondestructive method for assessing a bond-line, fasteners will remain. Structural failure can be catastrophic and is a constant reminder that the small faults can propagate to large and highly impactful results. Figure 8.4 shows the animated reconstruction of the in-flight failure in November 2007 (caused by fatigue cracking of a forward fuselage longer on). The crash occurred near Salem, MO, USA, during a training exercise. Figure 8.5 shows where the fatigue crack occurred.

As a final note, it is a common misconception that lightly loaded structures do not require rigorous surface preparation. The consequences of failure should be based on

FIGURE 8.5 Place where the fatigue crack occurred.

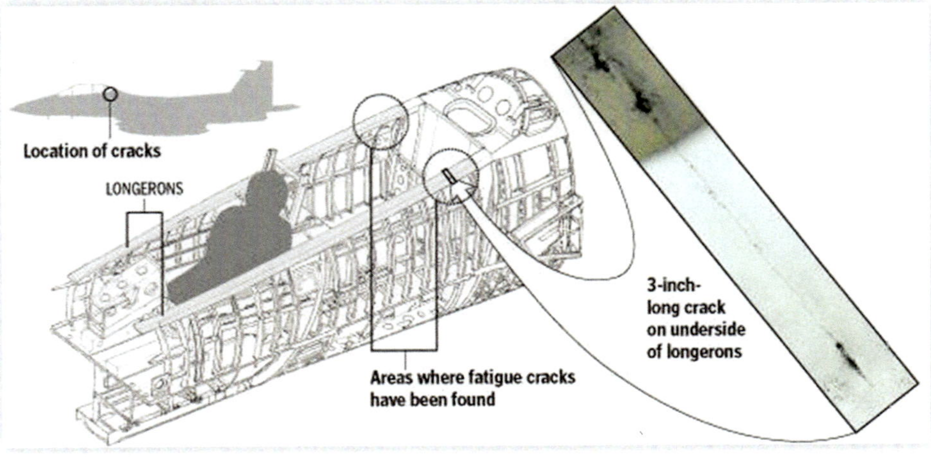

© SAE International/George Bullen

inadequate surface preparation, which will have poor environmental resistance and eventually the bond will fail. If there is any risk to the safety of flight as a consequence of the bond failure, a rigorous surface preparation should be undertaken. Surface preparation is critical to the safety of bonded structures. For lightly loaded structures, the requirement for surface preparation must consider the consequences of failure.

Many misconceptions involve the selection of materials for adhesive bonding. A common response to a bonded-joint failure is to change the adhesive, when the failure is more attributable to processes such as insufficient surface preparation techniques. A careful examination of the failure surface to identify the failure mode is essential to correct management of adhesive bond failures [8-14].

References

8-1. Kinlock, A. J. Adhesion and Adhesives. Chapman and Hall: New York, NY, USA. 1987. p. 78.

8-2. Arnott, D. R., et al. "Research Underpinning the Adherend Surface Preparation Aspects of the RAAF Engineering Standard C5033," International Aerospace Congress, Sydney, Australia, February 25–28, 1997, p. 41.

8-3. Comyn, J. Adhesion Science. Royal Society of Chemistry: Cambridge, U.K, 1997.

8-4. Bonk, R. B., Osterndorf, J. F., Ambrosio, A. M., Pettenger, B. L., and Froelich, K. A. "Evaluation of Adhesives for Adhering Carbon/Epoxy Composites to Various Metallic Substrates," Proceedings of the 41st International SAMPE Symposium, Anaheim, CA, USA, March 24–28, 1996, pp. 1472–1485.

8-5. Davis, M. J. "Deficiencies in Regulations for Certification and Continuing Airworthiness of Bonded Structures," International Aerospace Congress, Sydney, Australia, February 25–28, 1997.

8-6. Airframe and Power plant Mechanics Airframe Handbook, FAA AC65-15A, January 1, 1976. https://www.faa.gov/regulations_policies/advisory_circulars/index.cfm/go/document.information/documentID/23018. Accessed March 23, 2016.

8-7. Davis, M. J. "A Call for Minimum Standards in Design and Application Technology for Bonded Structural Repairs," Proceedings of the Symposium on Composite Repair of Aircraft Structures, Vancouver, BC, Canada, August 1995, pp. 4-1–4-15.

8-8. Hart-Smith, L. J. "An Engineer's Viewpoint on Design and Analysis of Aircraft Structural Joints," McDonnell Douglas Paper MDC 91K0067, also published in Proceedings of the Institute of Mechanical Engineers, Part G, Journal of Aerospace Engineering, vol. 209, 1995, pp. 226–238.

8-9. Davis, M. J., and Bond, D. A. "Towards a Certification Standard for Adhesive Bonded Structures and Repairs," The Symposium of the International Committee on Aircraft Fatigue ICAF-99, Seattle, WA, USA, July 12–16, 1999.

8-10. Hart-Smith, L. J. "An Engineer's Viewpoint on Design and Analysis of Aircraft Structural Joints," McDonnell Douglas Paper MDC 91K0067, also published in Proceedings of the Institute of Mechanical Engineers, Part G, Journal of Aerospace Engineering, vol. 209, 1995, pp. 226–238.

8-11. Thrall, E. W., Jr. "Bonded Joints and Preparation for Bonding," Proceedings of the AGARD-NATO Lecture Series, no. 102, 1979, pp. 5-1–5-89.

8-12. Hart-Smith, L. J., Redmond, G., and Davis, M. J. "The Curse of the Nylon Peel Ply," 41st International SAMPE Symposium, Anaheim, CA, USA, March 25–28, 1996.

8-13. Hart-Smith, L. J., Ochsner, R. W., and Radecky, R. L. "Surface Preparation of Fibrous Composites for Adhesive Bonding," Douglas Service, First Quarter Report, 1984, pp. 12–22.

8-14. Davis, M. J. and Bond, D. A. "The Importance of Failure Mode Identification in Adhesive Bonded Aircraft Structures and Repairs," The International Conference on Composite Materials ICCM-12, Paris, France, July 5–9, 1999.

CHAPTER 9

Operation

The logic of validation allows us to move between the two limits of dogmatism and skepticism.

—**Paul Ricoeur**

9.1 Introduction

The use of controlled, localized, stress waves in materials and bonded joints offers new opportunities for the characterization of structures. Bonded joints in particular benefit from this application because there is no nondestructive inspection method to measure bond strength. Laser bond inspection using a high-peak-power short-pulse laser can perform a localized proof test of a bond. The method can be applied nondestructively to strong bonds but will fail a weak bond, creating an internal dis-bond detectable by the laser bond inspection system itself or by post-test ultrasonic inspection. To date, many tests have shown the method to be sensitive to weak bonds in carbon fiber reinforced polymer composite structure bonds created by poor adhesive mixing, improper surface preparation, or contamination.

Bonding offers considerable opportunity to reduce manufacturing cost and improve structural performance [9-1]–[9-3]. However, the bonding of aerospace structures continues to meet with concerns from certification agencies because of past experiences with bond-line failures. Those failures have been predominately attributed to material and surface preparation issues [9-4]. Nondestructive evaluation (NDE) plays a key role in the evaluation of bonded joints. NDE methods, however, cannot directly measure the key parameter of interest, which is adhesive strength. NDE does measure characteristics that under certain conditions can indicate a potential loss of strength, or a weakened bond relative to a strong bond [9-5]–[9-13]. The literature has examples of correlations between sophisticated data extraction from measurements and weak bonds. However, these techniques have not been taken to the next level of implementation on real bonded structures. This is likely because of the concern that laboratory tests are subject to the specific method of weak bond creation and are not necessarily sensitive to all factors that could weaken an adhesive interface. Inserts, voids, thickness variation, and even adhesive changes can have distinctive features that can be detected and correlated to a mechanical strength change. On the other hand, interfaces that are weakened by contamination or improper surface preparation are the most difficult to sense because there often are no macroscopic feature effects.

The true strength of a bond is measured in the plastic regime of material behavior and, therefore, is not a parameter that would be directly measured by nondestructive (elastic) measurements. An approach of interest, therefore, is to create internal stress waves that can locally proof test an adhesive interface and be applied as part of the NDE process. Mechanical waves of sufficient energy level can be created that will locally apply tension at a bond interface. Weak bonds will fail under a design test load, while strong bonds will survive.

9.2 Laser-Generated Stress Waves

Extensive experimental development, supported by 1-D and 2-D hydrodynamic code simulations, has demonstrated that the strength of adhesive bonds can be tested using calibrated weak shock waves (stress waves). Gupta et al. [9-14], Yuan and Gupta [9-15], and Boustie et al. [9-16] demonstrated laser shock methods for surface spallation. Bossi et al. [9-17] demonstrated that internal bonds could be evaluated with shock waves. High-peak-power lasers have been used to generate the internal stress waves suitable for bond evaluation [9-18]–[9-20]. Modeling of the method has shown that the laser beam shape results in controlled, very localized testing of bond strength. **Figure 9.1** illustrates the fundamental process of laser-generated stress wave evolution in a substrate. The energy delivered by the laser is absorbed in a sacrificial overlay such as black paint at the incident surface of the sample. No sample heating takes place, and there is no surface damage. Vaporization of some elements of the overlay produces locally high pressure that is enhanced by inclusion of a transparent overlay layer—such as a water film—that briefly confines vapor expansion. This pressure pulse duration can be tailored to a few hundreds of nanoseconds. The magnitude of the pressure is a function of the laser input irradiance, which facilitates generation of calibrated stress waves. The shock from the laser pulse produces a compression wave that propagates to the back surface of the sample, where it reflects back in tension. In the 1-D approximation, it is this tensile wave propagating through the sample that provides the proof-test loading. When the wave arrives back at the front surface, it reflects into compression again and the cycle repeats. The lower part of Figure 9.1 illustrates the time evolution of the stress wave for a simplified 1-D case in which the laser beam diameter is at least several times larger than the sample thickness. The actual 2-D case of practical interest is considerably more complex, but the basic

FIGURE 9.1 1-D approximation for laser-based shockwave generation and stress pulse propagation in a solid slab.

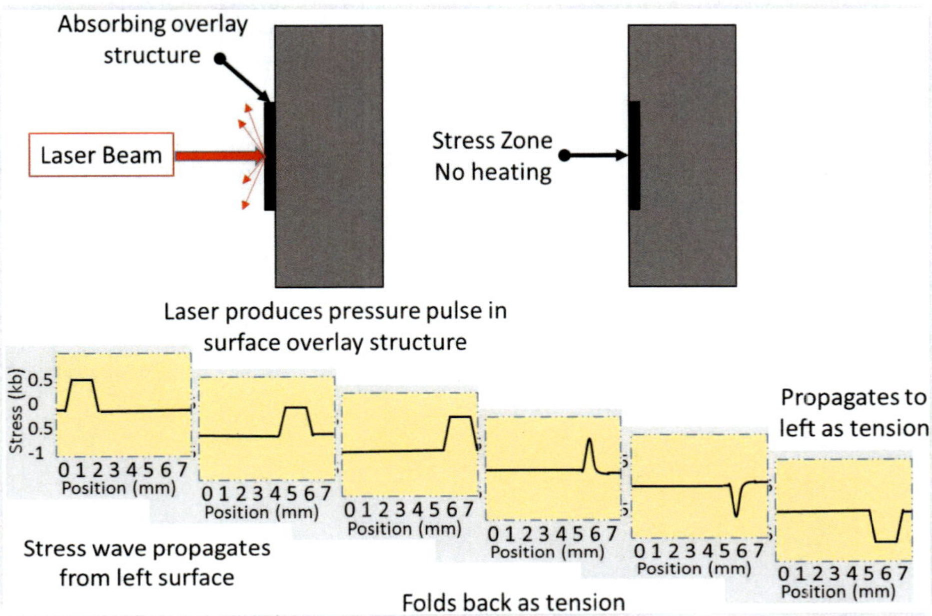

principles are the same. As the wave reflects from a free surface, it folds back through the incoming wave and relieves stress to zero in the overlap region. Net tension appears only after the reflected wave clears the tail of the incident wave. Thus, in the simple 1-D approximation, without attenuation, all portions of the sample experience equal tension duration, except for an excluded region near each surface where the duration of tension decreases linearly to zero with proximity to the surface. Dynamic failure is frequently dependent on both stress and duration. This suggests that the near surface regions may not be able to be inspected at the same level as the sample bulk.

9.3 Laser Bond Inspection Development

LSP Technologies, Inc. (LSPT), Dublin, OH, USA, employed a suitable Nd:glass laser (1054-nm wavelength) for development of a laser bond inspection technology. The LSPT laser produces 45 J/pulse with very high reproducibility. The laser was tailored to provide Gaussian-like pulse widths of 70–300 ns. **Figure 9.2** shows the beam power profile and the beam spatial distribution that is relatively flat topped. Beam diameter is selected as a compromise between the need to have a large area for planar wave generation and a reasonable-sized beam for the inspection of small zones in the object; 1-cm-diameter beams provide a suitable compromise. The use of large diameter (> several millimeters), broad pulse (> 50 ns) laser beams in this approach generates suitable internal stress for evaluation of internal bonds and avoids surface spallation.

FIGURE 9.2 Laser power pulse profile and laser spatial distribution map.

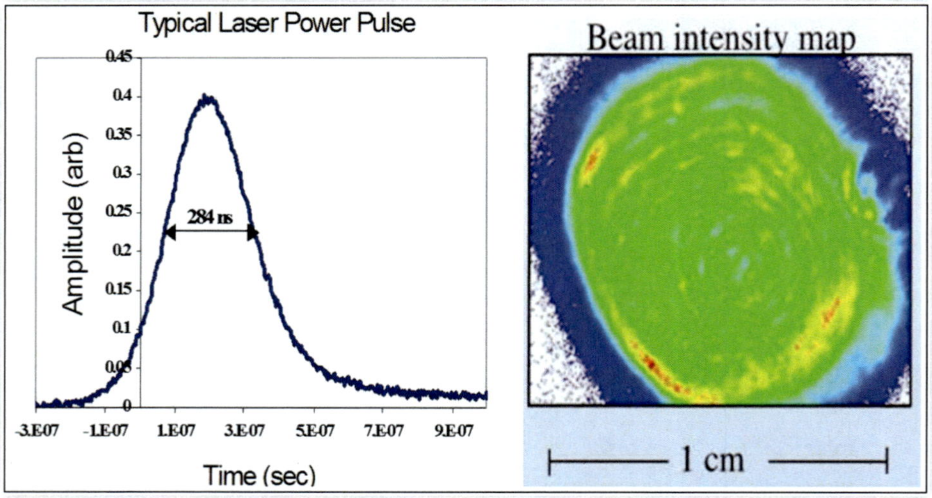

Numerical modeling of the process using CTH[1], a shock wave code, developed by Sandia National Laboratory, showed that having a large-diameter beam is important for development of adequate on-axis tension. If the beam diameter is less than roughly the part thickness, relief waves from the beam edge reduce the on-axis stress significantly. In general, waves originating at the beam edge produce surface velocity histories that are more complicated than the relatively simple 1-D response obtained when the beam is at least several times larger than the part thickness. This is illustrated in **Figure 9.3**, which shows the back-surface velocities measured for four aluminum plates, of variable thickness, subjected to a 19-mm-diameter beam. The thinnest plate shows the expected 1-D response for the first couple of microseconds. That is, the velocity increases to a peak, corresponding to the arrival of the main compressive shock at the back surface, and then decreases to nearly zero as the pulse reflects in tension.[2] Roughly one microsecond later, the tensile wave has reached the front surface, reflected in compression, and returned to the back surface to produce the second peak. Subsequently, waves from the beam edge begin to complicate the velocity history. For thicker plates, these edge effects appear relatively sooner and give the velocity response a markedly different character. For example, the velocity swings negative immediately after the first positive pulse. Note that in the figure the timescales were adjusted so that the first peak arrives at the same apparent time in the graph to allow easy comparison of the waves.

[1] CTH is a multi-material, large deformation, strong shock wave, solid mechanics code developed at Sandia National Laboratories. It has models for multi-phase, elastic, viscoplastic, porous and explosive materials. Three-dimensional rectangular meshes, two-dimensional rectangular and cylindrical meshes, and one-dimensional rectilinear, cylindrical, and spherical meshes are available. CTH has adaptive mesh refinement and uses second-order accurate numerical methods to reduce dispersion and dissipation to produce accurate and efficient results. CTH provides an end-to-end simulation solution including visualization support and runs on Linux, Windows, Macintosh and UNIX workstations, Beowulf clusters and massively parallel supercomputers. Please note that CTH is government owned export controlled software and can only be distributed to U.S. citizens or to non-citizens under an approved U.S. Dept. of State export license.

[2] The velocity does not return to zero because the shine time of the laser was slightly greater than the round-trip transit time through the 3-mm plate.

FIGURE 9.3 Back-surface velocity for a 19-mm diameter laser pulse as a function of aluminum specimen thickness.

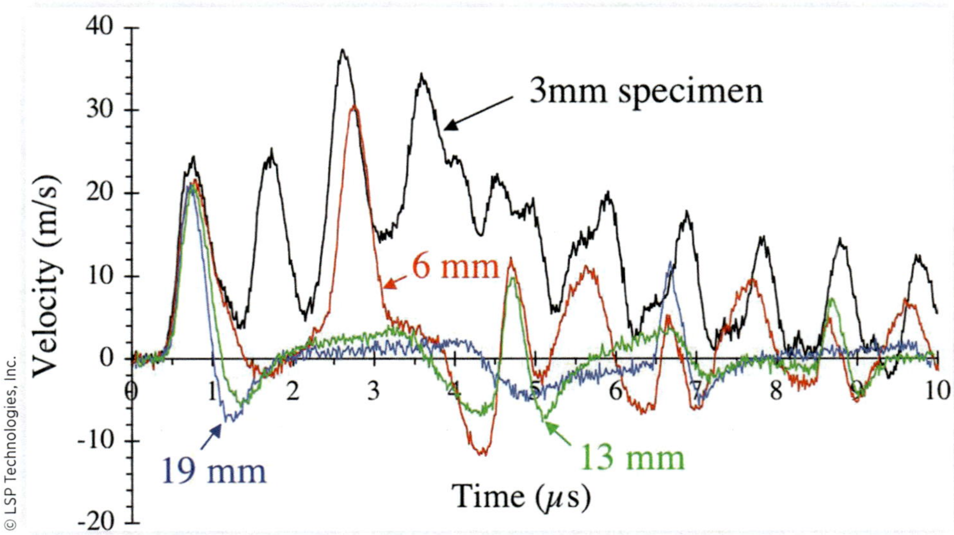

2-D CTH simulations explain the complex nature of the velocity histories measured in the experiments. **Figure 9.4** shows the result of one such simulation. The figure shows the velocity history for an on-axis point at the back surface of a 19-mm-thick plate. The inset shows a snapshot from the simulation showing the normal component of stress within the sample. Yellow, orange, and red colors represent compressive stress, blue represents tension, and greens are nearly zero. At the time of the snapshot, the main compressive shock is about to arrive at the back surface of the specimen. This compression produces the first positive peak in the back-surface velocity history. If the beam diameter was very large, this main compression is all that would appear, and the simple velocity history shown for the thinnest

FIGURE 9.4 CTH 2-D simulation of the stress distribution in a 19-mm-thick aluminum sample for a 19-mm-diameter laser beam.

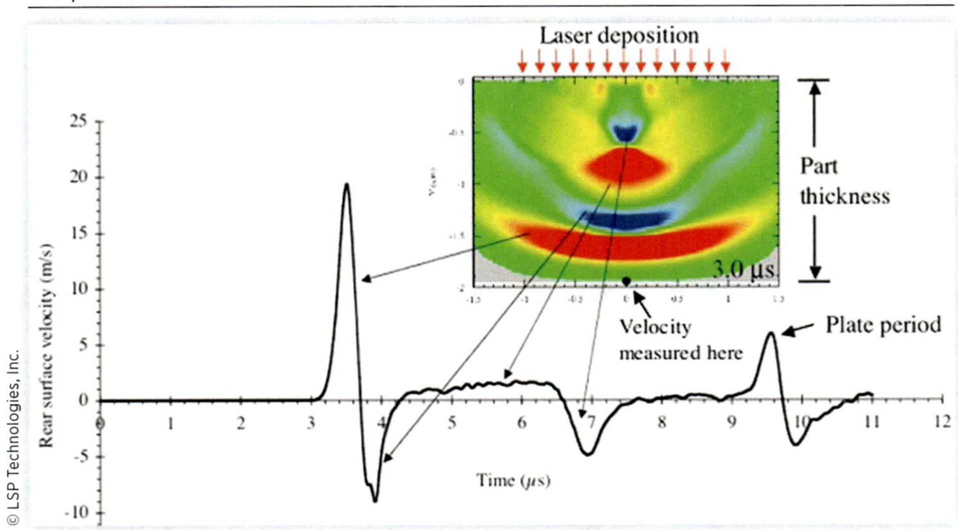

FIGURE 9.5 Comparison of 2-D hydrodynamic simulation and experimental measurement of surface velocity as a function of sample thickness.

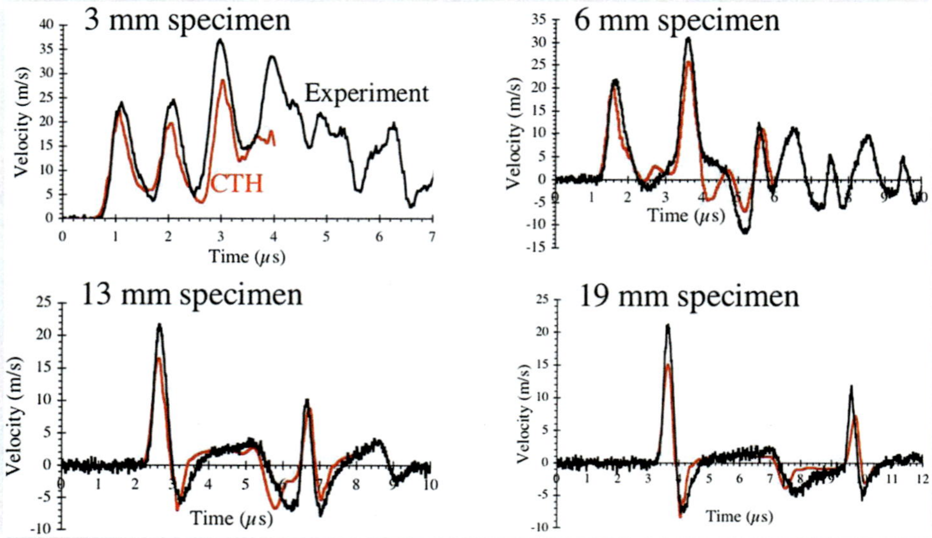

sample in Figure 9.3 would result. Instead, waves from the beam edge produce a complex series of alternating tension and compression waves behind the main shock. The arrows in Figure 9.4 show the correspondence between these wave and the various negative and positive features in the back-surface velocity history. The alternating compression and tension features produce the positive, negative forms of the velocity history. The CTH simulations compare very well with the experiments, as shown in **Figure 9.5**. The simulations provide a means to optimize the stress distribution in a part by the selection of beam diameter or other pulse characteristics as a function of part thickness and the desired location for peak tension.

Surface velocity is the critical characteristic for this stress wave measurement, from which stress can be deduced. The stress is calculated as $\sigma = pcu$, where p is density, c is longitudinal sound speed, and u is measured surface velocity. A Valyn Velocity Interferometer System for Any Reflector (VISAR)[3] is typically used on the back surface of the test object to measure surface velocity. The VISAR bandwidth is better than 200 MHz, resulting in highly reproducible and accurate measurements of the sample surface velocity. An electromagnetic acoustic transducer (EMAT) has also been developed as a surface velocity-sensing instrument. The EMAT can be used either on the front or back surface and is specially designed for each application.

[3] The Velocity Interferometer System for Any Reflector (VISAR) was developed by Barker and Hollenbach at Sandia National Laboratories in 1972. The technique works by using an interferometer to measure the minute Doppler shift in light frequency given to a laser beam as it is reflected from a moving specimen surface. The Doppler shift produces light fringes in the interferometer, and the number of fringes is proportional to the surface velocity. VISAR data consists of recordings of the light intensity outputs from the VISAR interferometer as a function of time, in which each complete oscillation of the light intensity corresponds to one light fringe. VISAR data reduction consists of counting the number of light fringes up to any given time, and multiplying by the Velocity-Per-Fringe constant of the interferometer. Repeating this procedure for each time increment results in a complete velocity vs. time record of the specimen surface. Using a data reduction program, the fringe count can usually be determined to ± 2% of a fringe or better.

FIGURE 9.6 Peak stress as a function of laser fluence and overlay material. PTBT—packing tape, black tape, WBPSTS—water, black paint, stainless steel, and WBT—water, black tape.

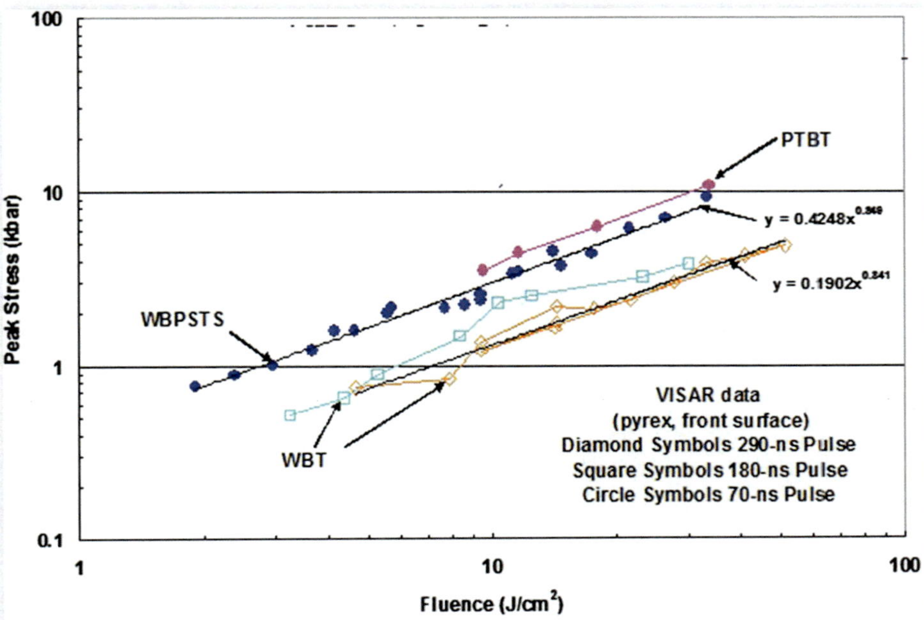

Measurements of surface velocity, converted to peak stress, are shown in Figure 9.6 for several different sacrificial overlay layers. The figure shows a predictable relationship between stress generation and laser fluence over a wide range, more than sufficient for proof testing. There is a difference in the level of stress as a function of the overlay selection and the pulse width. It is observed that a higher peak stress occurs for short pulses at a given fluence than for longer ones. In terms of bond-line delamination requirements, wider pulses compensate for lower peak stress so that, to a reasonable approximation, failure fluence requirements are similar (but not identical) for different pulse lengths.

9.4 Laser Bond Inspection Test Results

Experimental measurements of bonded joints have been performed on a variety of samples. The primary effort has been on composite-to-composite bonds using room temperature paste adhesives or hot-film adhesive bonds. A few experiments with aluminum-to-aluminum and aluminum-to-composite have been tried. For composite bonds, the adherent thickness has ranged from approximately 2-mm (0.080 in.) thickness up to over 20-mm (0.8 in.) thickness with a total bond thickness of 25 mm (1 in.). Most of the effort has used bonded material coupons in the 6- to 10-mm (0.25 to 0.4 in.) thickness range.

To validate that the method can detect weak bonds, it is necessary to construct samples that have a range of bond strengths from very weak to full strength for the process. The methods employed to construct weak bonds have included mis-mixing the epoxy adhesive, inappropriate surface preparation, and controlled contamination of the surfaces. Bonds were made with Loctite™ Hysol EA 9394 epoxy mixed in the standard hardener-to-resin ratio (B/A) of 0.17, with B/A = 0.05 and with B/A = 0.03. These mixtures provide full static strength, 70% strength, and 50% strength bonds, respectively. Various surface preparations were also employed. They included grit blast (GB), "over-" or double-GB (OGB), sanded and solvent cleaned (S), and surfaces roughened with Scotchbrite, as tooled with solvent wipe (Sol), and peel-ply (PP) removal only. PP variations have included silicon release blue (SRB), nylon, and polyester PP materials. Surfaces were also prepared with 500 and 5000 ng/cm^2 of Frekote™ 1711 contamination to test for weak joints. Bond-lines were prepared with nominal thickness (~0.5 mm) and with thickness steps up to 3 mm. Baseline samples used 20 plies bonded to 20 plies with thickness/ply = 0.21 mm. Additional laminate pairings were also studied using 10, 16, 20, 30, and 46 plies in all combinations to evaluate the range performance of bond-line response. Individual laminates (up to 60 plies) and freestanding adhesive samples were exposed to laser-induced shock to establish the dynamic failure threshold for the constituents of the bonded samples.

Prior work had demonstrated rear surface spallation of thin laminate material exposed to laser-generated shockwaves [9-14], [9-15]. An important initial question was whether bond lines could be evaluated under conditions that would not first produce failure in the laminate, because tensile stress appears first near the free surface. Static flatwise tension tests of well-made bonded laminates show that laminates and adhesive static strengths are similar. Since shockwaves attenuate with distance owing to geometric and bulk loss processes, it is expected that higher tensile stress will be produced at the back surface of laminate pairs than will be required to produce tensile failure of bond lines near the front surface of the parts. Thus, a 10ply/46 ply combination (with 10 ply as the laminate closest to the laser deposition) might not fail in the bond-line before delamination of the 46-ply element. During more than 3000 experiments, it was consistently demonstrated that bond-line failure is induced prior to laminate failure onset for most pairings of the laminates listed above. Laminate failure, simultaneous with bond-line failure, was observed for the 10/46-ply combination in several experiments, but only with full strength bond-lines having ideal surface preparation.

Figure 9.7 shows a typical setup for testing. The pulsed laser beam was projected normal to the sample surface following the large arrow in the bottom right of the figure to form a 1-cm-diameter spot at projected cross hairs on the sample. It should be noted, however, that normal incidence is not a requirement since the beam arrival time is essentially instantaneous over the spot, and impulse generation is normal to the surface. An overlay structure is visible in the figure, consisting, in this case, of clear packing tape on black vinyl tape. A water jet flowing over black acrylic spray paint was actually used for much of the work. More complex overlays have been used to tailor shock wave generation in the samples. All produced similar sample response when appropriate adjustments were made to the delivered laser fluence and, in some cases, the laser pulse length.

In the present context, range is considered as the total wave travel distance from insertion as compression, to the sample back and then as tension to the bond-line. The range dependence of bond-line failure is important because it establishes the total thickness and laminate pairings that can be successfully tested, and establishes a requirement for laser performance capability. In addition, it may provide an indication of whether the input stress rate is appropriate to the application. **Figure 9.8** shows a plot of fluence required to

FIGURE 9.7 Experimental arrangement.

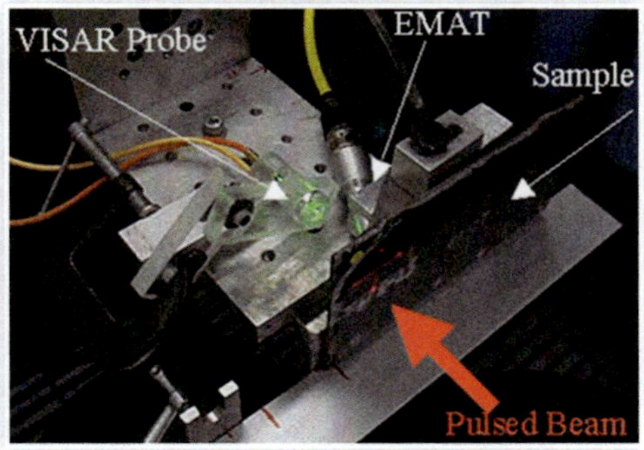

FIGURE 9.8 Threshold fluence for bond-line failure versus wave travel distance to produce bond-line tension.

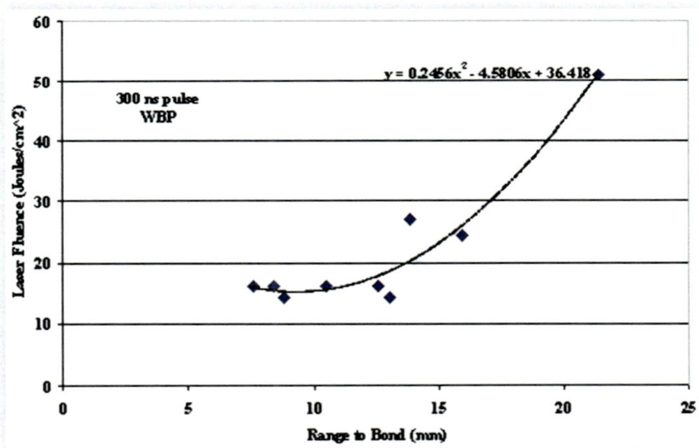

fail a well-made bond in Cytec Fibertite™ T300 934 graphite epoxy/EA9394 as a function of range. Representative laminate pairings are indicated in the figure. It is expected that the fluence requirement should increase with range to compensate for attenuation and geometric spreading of the stress wave. The figure is for a 1-cm-diameter laser spot and shows a rapid increase in required fluence when the range to the bond-line is more than about 3 spot radii (approximately 15 mm in Figure 9.8) from the laser input surface. Similar curves for all other bond preparations fall below the one shown. Failure of the back laminate lies above this curve in most cases, although some measurements indicate that laminate failure occurs at nearly the same fluence as the bond-line for the 10/46-ply sample at the right. The observed scatter is principally because only a few measurements contribute to this curve and many experiments are required for good statistics. Many measurements have been completed on "standard" 20/20 samples that show a tight grouping of outcomes.

Figures 9.9–9.12 show results from laser bond testing of samples with variable strength. The graphs plot laser fluence on the horizontal axis versus a damage parameter. The damage parameter is 0 when no detectable damage can be found by ultrasonic inspection following the testing. A value of 1 is assigned when the bond is clearly delaminated. A value of 0.5 is assigned to a perceptible change in the UT signal. The charts show that as the fluence delivered increases, damage is detected at the weak interface prior to stronger interfaces. The figures indicate that criteria for proof testing of bonds can be established that will fail weak bonds at specified laser fluence levels but not strong bonds. Figure 9.9 compares three methods for adhesive strength of EA 9394 two-part epoxy adhesive on grit blasted composite surfaces. As the fluence is increased, the weak bond mixture fails around 6 J/cm^2. At a higher fluence the medium strength fails, while the regular strong mixed adhesive bond does not fail until 16 J/cm^2. The data show clear discrimination

FIGURE 9.9 Threshold for bond-line failure for standard EA9394 and weak mixture preparation.

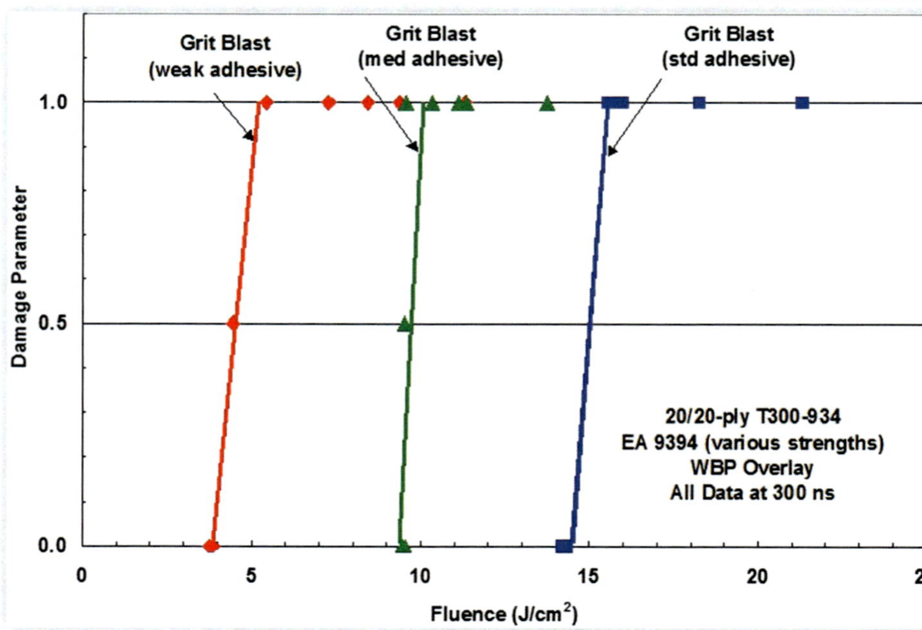

FIGURE 9.10 Threshold for bond-line failure for various strength EA9394 and standard grit blast surface preparation.

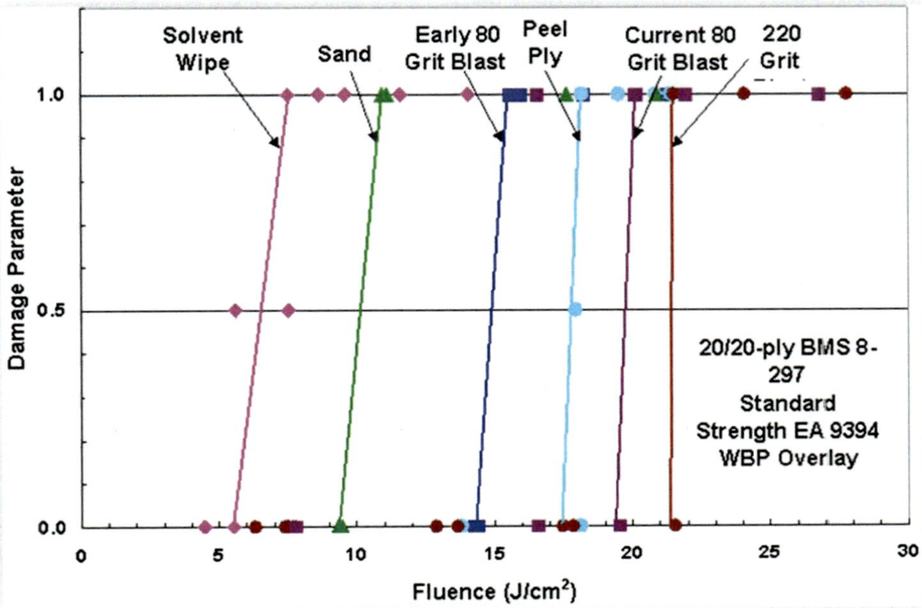

FIGURE 9.11 Threshold for bond-line failure for standard several PP surface preparations.

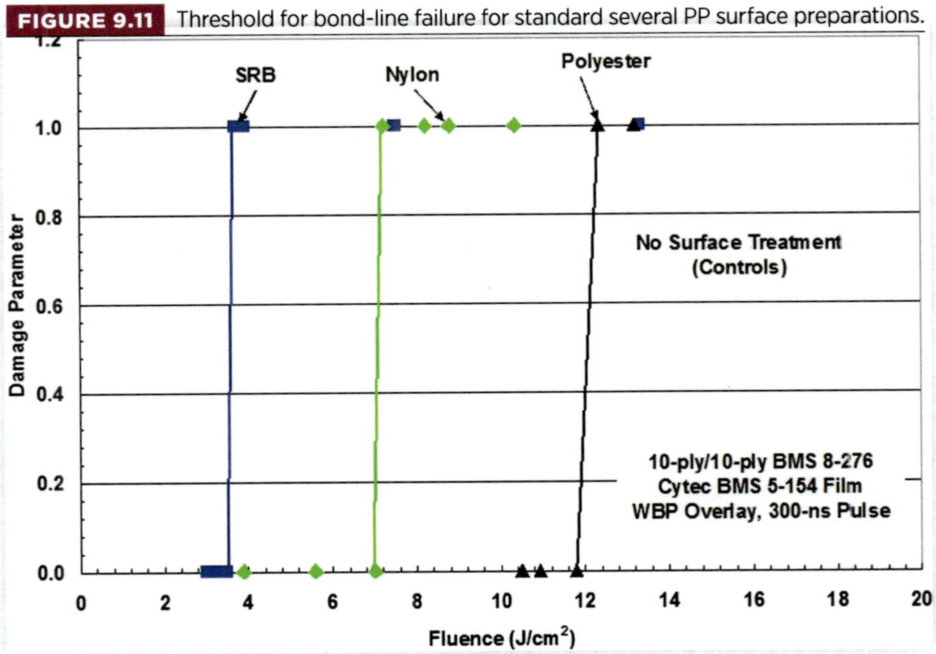

of weak-, medium-, and full-strength adhesive. Mechanical tests of the mixture indicated that the weak mixture is about 50% of the full strength mixture and the medium mixture is about 75%. Figure 9.10 shows a study of six types of surface preparations. As the fluence is increased, the failure level of each method is identified. The results are in agreement with mechanical tests. Figure 9.11 shows a comparison of the three types

FIGURE 9.12 Threshold for bond-line failure for various levels of contamination of Frekote™ 1711 on GB surface preparation with EA9394 paste adhesive.

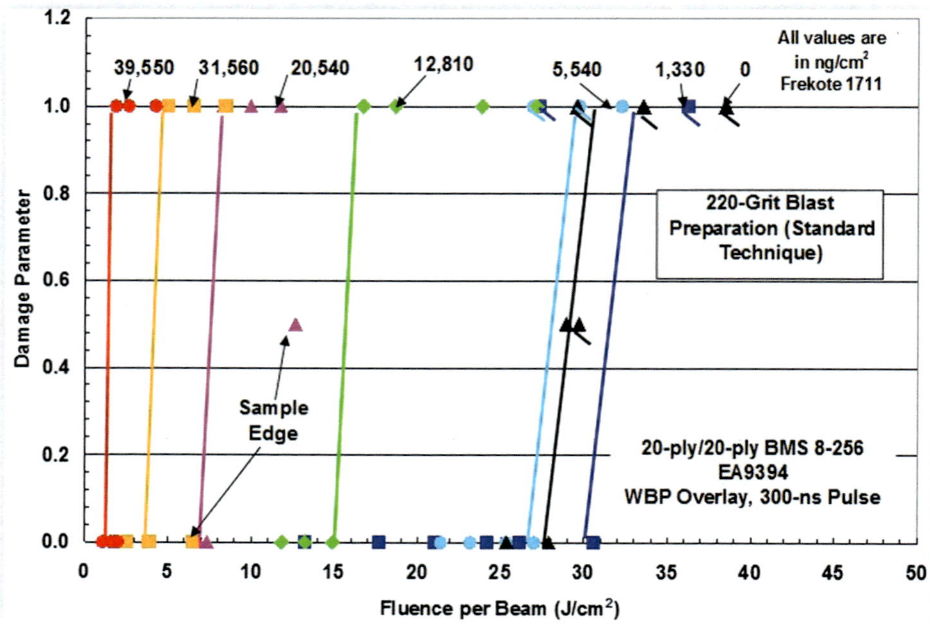

of PP surface preparations. The SRB is the weakest (as it is designed to be), resulting in a bond failure of around 4 J/ cm². The polyester PP is the strongest at 12 J/cm². Figure 9.12 illustrates detection of Frekote™ 1711 contamination levels on a grit blasted surface with standard EA9394 mixture. The Frekote™ concentrations are estimated between 0 and 40,000 ng/cm². Significant loss of strength occurs around the 10,000-ng/cm² level.

Figure 9.13 shows a comparison between the double cantilever beam strain energy release rate ($G1C_{1C}$) versus the applied laser fluence for laser bond inspection bond-line failure. The results show a reasonable correlation. **Figure 9.14** addresses a critical question for this method of bond-line proof testing, i.e., does the laser inspection technique itself induce progressive damage to the sample? The figure shows the dependence of bond-line failure on the number of pulses delivered to the sample at a given laser fluence. The results are plotted as a function of measured rear-surface velocity for 20/20-ply samples, because velocity is proportional to stress. These data are from early tests, obtained with 70-ns stress pulses, and employed overlays with stress conditioning layers. Similar behavior is observed with 300-ns pulses using a WBP overlay. Results show that bond-line delamination occurs at lower input fluence/velocity if repeated pulses are applied. However, they also indicate that the behavior asymptotes to no-damage upon application of repeated pulses within 60%–70% of single-pulse threshold conditions. This outcome shows that the bond-lines can be easily subjected to loads of up to 50% of dynamic single-pulse ultimate, for the purposes of proof testing. Since multiple pulses are required to produce detectable damage at higher loads, single-pulse proof testing can probably be safely conducted at a higher percentage of ultimate. It is to be noted that mechanical tests of laser-shocked samples have never exhibited a reduction in static strength in flatwise tension, double lap shear, or 3-point bending, even when damage detectable with UT existed at the exposed site.

FIGURE 9.13 Strain energy release rate versus laser fluence for weak bond testing.

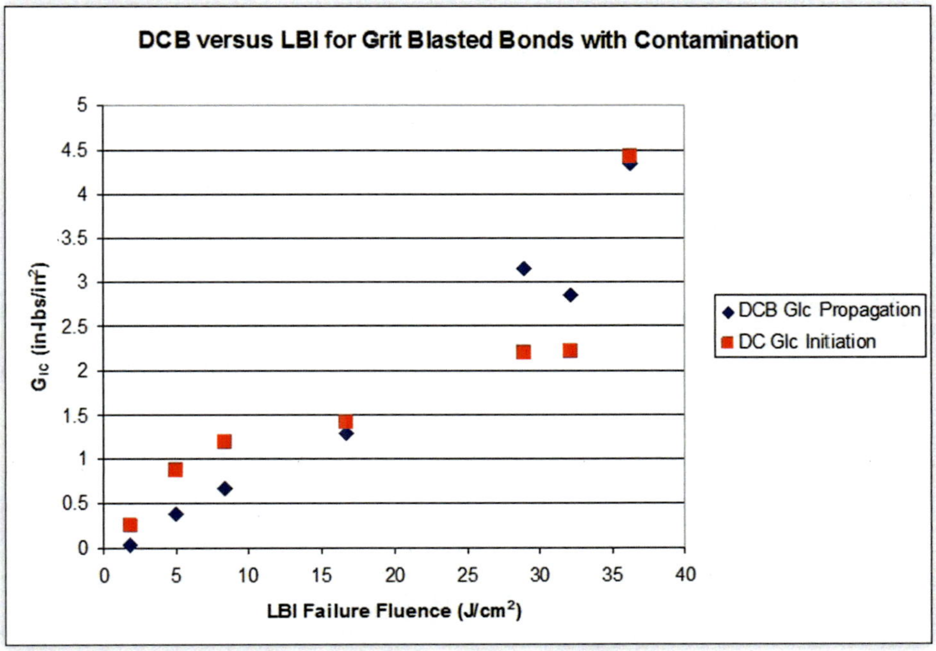

FIGURE 9.14 Progressive damage effects with two overlays and 70-ns pulse.

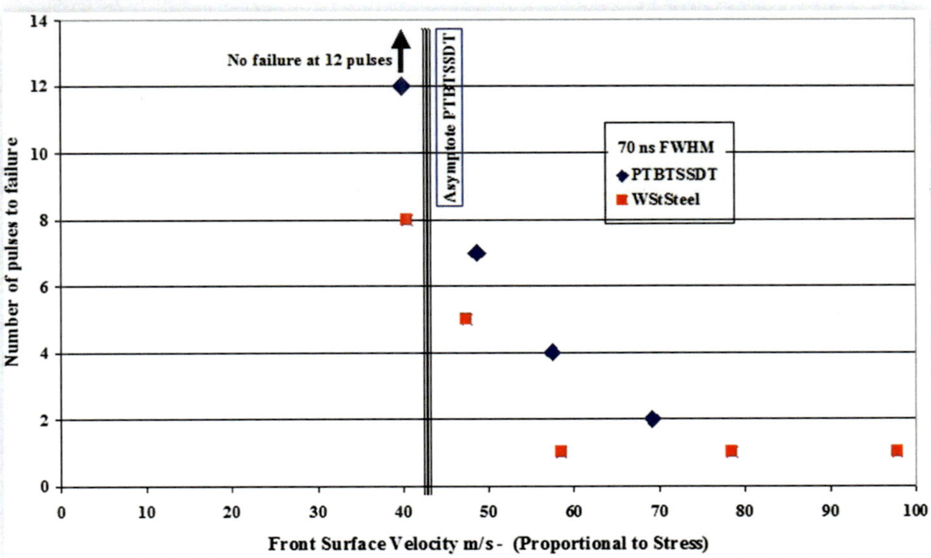

9.5 Laser Bond Inspection System

Based on the success of the experimental studies, a laser bond inspection device (LBID) has been developed under AF SBIR funding. The system consists of a high-peak-power pulsed laser, an articulated beam delivery arm, a contact process head, and computer control. **Figure 9.15** shows the size and configuration for the laser unit on a transportation carriage with a beam delivery arm.

Figure 9.16 shows a laser beam delivery arm attached to the inspection head that is applied to bonded structures during LBI testing. The head provides a vacuum to hold the system stable to the test part, a water confining layer for the pulse pressure generation, and an EMAT-based surface velocity measurement system. A computer workstation

FIGURE 9.15 Laser bond inspection device on transportation carriage with beam delivery arm.

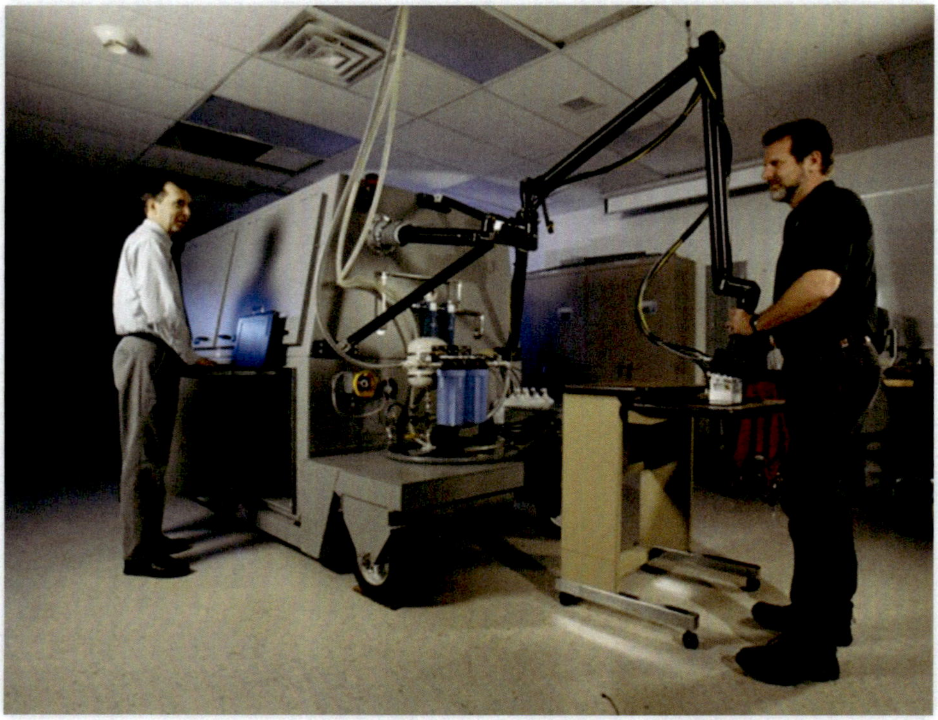

FIGURE 9.16 Articulated arm inspection head and base of head for beam delivery to bonded test structures.

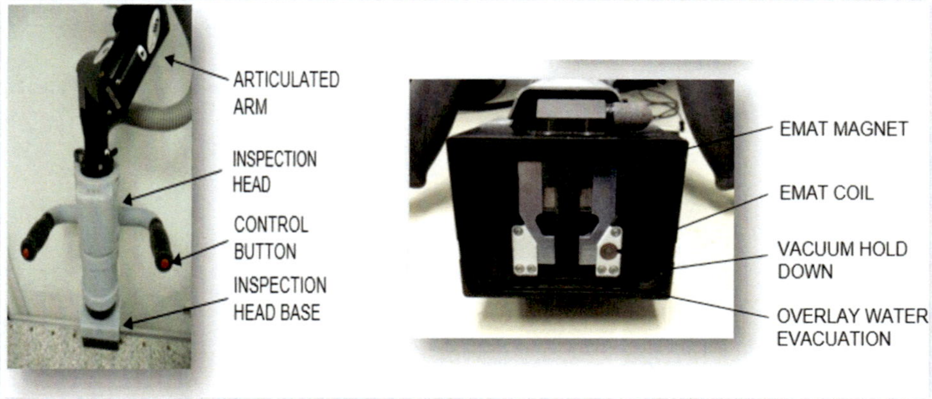

FIGURE 9.17 Computer control panel.

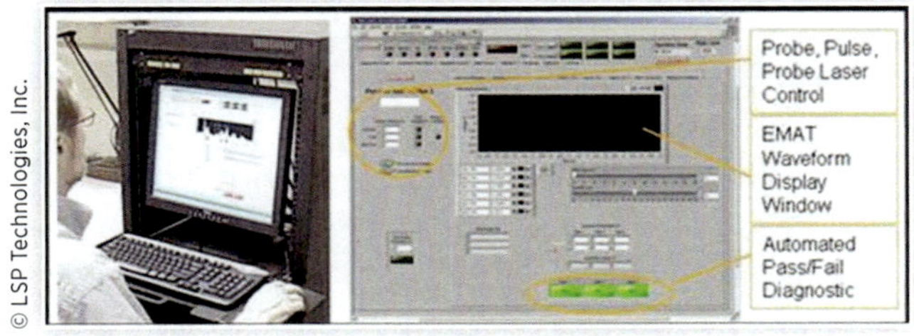

provides laser pulse control and diagnostics. **Figure 9.17** shows the computer control screen for a laboratory-based LBID. **Figure 9.18** shows the results for the testing of a weak and strong bond. In this scenario, the system delivers three pulses: 1) a low-fluence probe pulse, 2) a high-fluence test pulse, and 3) a second low-fluence post-test probe pulse. Comparing the surface velocity signals from the first and third pulses, the computer can determine whether the bond has been failed. Alternatively, the system may be operated with a single or a sequence of laser pulses followed by ultrasonic inspection to determine any bond-line damage associated with particular laser fluence level.

FIGURE 9.18 Computer analysis of signals from laser bond inspection test sequence of probe, test, and probe laser fluence to determine bond strength.

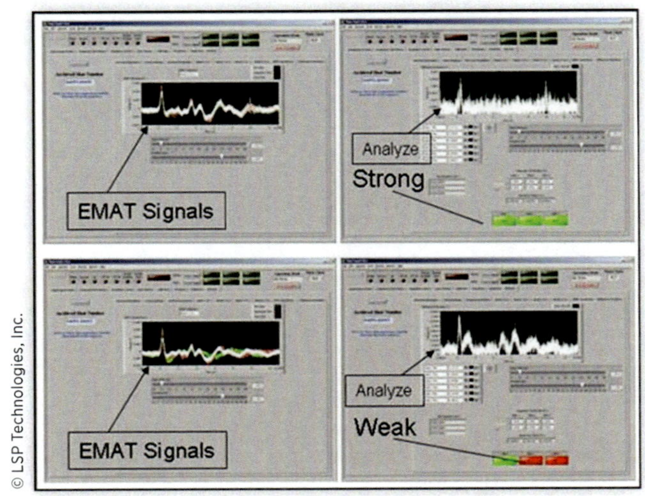

9.6 Summary

Extensive experimental development, supported by 1-D and 2-D hydrodynamic code simulations, has resulted in the development of a laser bond inspection test method for assessing bond strength. The method uses a high-peak-power pulsed laser to provide a localized dynamic proof test for composite-to-composite bonds that have a free back surface and suitable acoustic transmission properties. Sensitivity to bond quality variations such as surface preparation, adhesive mixing, and contamination have been demonstrated in repeatable tests.

However, much work needs to be done to develop complementary technologies that can be used to complete the suite of nondestructive testing of adhesively bonded structures. Verifying the integrity of an adhesive bond-line that joins two pieces of structure together is the first step toward providing the level of security needed to integrate an assembly without fasteners into a vehicle for widespread public use.

Acknowledgment

This effort was supported by the Composite Affordability Initiative, a USAF and industry consortium effort. Experimental testing was performed at LSP Technologies, Inc., Dublin, OH, USA. This chapter was reproduced from a paper with permission from David Sokol and David Lahrman, LSP Technologies, 6145 Scherers Place, Dublin, OH, USA (www.lsptechnologies.com).

References

9-1. Kinloch, A. J. Adhesion and Adhesives: Science and Technology. Chapman and Hall: London, UK. 1987.

9-2. Butler, B. "Composites Affordability Initiative Overview," Paper presented at SME Westec 2003, March 24, 2003.

9-3. Wanthal, S. "The Composites Affordability Initiative - Phase 3 Overview," Proceedings SAMPE 2006, November 2006.

9-4. Hart-Smith, L. J., Brown, D., and Wong, S. "Surface preparations for ensuring that the glue will stick in bonded composite structures," Paper MDC 93K0126, 10th DOD/NASA/FAA Conference on Fibrous Composites in Structural Design, November 1993.

9-5. Cawley, P., Pialucha, T., and Lowe, M. "A Comparison of Different Methods for the Detection of a Weak Adhesive/Adherend Interface in Bonded Joints," in Review of Progress in Quantitative Nondestructive Evaluation, Vol. 12, edited by D. O. Thompson and D. E. Chimenti. Plenum Press: New York, NY, USA. 1993. pp. 1531–1538.

9-6. Wegman, R., and Mitchell, J. R. "Evaluation of Adhesive Bond Strength by Nondestructive Testing," 21st International SAMPE Technical Conference, September 25–28, 1989, pp. 196–209.

9-7. Hagemaier, D. "Adhesive-Bonded Joints," in Adhesive Bonding of Aluminum Alloys, edited by E. W. Thrall and R. W. Shannon. Marcel Dekker Inc.: New York, NY, USA. 1985. pp. 337–423.

9-8. Adams, R. D., and Cawley, P. "A Review of Defect Types and Nondestructive Testing Techniques for Composites and Bonded Joints," NDT International. 21(4):208–222, 1988.

9-9. Rokhlin, S. I., Hefets, M., and Rosen, M. "An Ultrasonic Interface-Wave Method for Predicting the Strength of Adhesive Bonds," Journal of Applied Physics. 52(2):2847–2851, 1981.

9-10. Dickstein, P., Segal, Y., Segal, E., and Sinclar, A. N. "Statistical Pattern Recognition Techniques: A Sample Problem of Ultrasonic Determination of Interfacial Weakness in Adhesive Joints," Journal of Nondestructive Evaluation. 8(1):27–35, 1989.

9-11. Smith, R., Weise, V., and Dalton, R. "Potential Advanced Ultrasonic Methods for Detection of Weak Adhesion," Proceedings of NDT 2003, 42nd Annual Conference of the British Institute for NDT, 2003.

9-12. Rokhlin, S. I., Adler, L., and Hamilton, A. "Quality Assurance of Composite Bonding Processes with Angle Bean Ultrasonic Spectroscopy (ABUS)," Proceedings SAMPE, Long Beach, CA, USA, 2005.

9-13. Roach, D., and DiMambro, J. "Enhanced Inspection Methods to Characterize Bonded Joints: Moving Beyond Flaw Detection to Quantify Adhesive Strength," Air Transport Association NDT Forum, Ft. Worth, TX, USA, October 2006.

9-14. Gupta, V., Yuan, J., and Pronin, A. "Experimental Techniques in the Dynamics of Deformable Solids," ASME, AMD-165, pp. 61–70, 1993.

9-15. Yuan, J., and Gupta, V. "Measurement of Interface Strength by the Modified Laser Spallation Technique," Journal of Applied Physics. 74(4):2388–2410, 1993.

9-16. Boustie, M., Berthe, L., de Resseguier, T., and Arrigoni, M. "Laser Shock Waves: Fundamentals and Applications," 1st International Symposium on Laser Ultrasonics: Science, Technology and Applications, July 16–18, 2008, Montreal, Canada.

9-17. Bossi, R. H., Housen, K. R., and Shepherd, W. B. "Using Shock Loads to Measure Bonded Joint Strength," Materials Evaluation. 60(11):1333–1338, 2002.

9-18. Bossi, R. H., Housen, K. R., and Shepherd, W. B. "Application of Stress Waves to Bond Inspection," SAMPE, Long Beach, CA, USA, May 16–20, 2004.

9-19. Bossi, R., Housen, K., and Walters, C. "Laser Bond Inspection Device for Composites: Has the Holy Grail Been Found?" Nondestructive Testing Information Analysis Center. 30(2):1–4, 2005.

9-20. Arrigoni, M., Kruger, S., Bouin, A., Levesque, D., Lord, M., and Monchalin, J. "The Use of Laser Doppler Interferometry Based on a Fabry-Perot Etalon for Shock Adhesion Test Applied to Adhesively Bonded Materials," 1st International Symposium on Laser Ultrasonics: Science, Technology and Applications, July 16–18, 2008, Montreal, Canada.

CHAPTER 10

Extensibility

Why, sometimes I have believed as many as six impossible things before breakfast.

—Lewis Carroll

10.1 Introduction

Note that in Chapter 7, Introduction to Section Two, the process change from a fastened structure to an adhesively bonded structure is risky and has to include enhanced discipline in the manufacture of the parts to be bonded together, precise and consistently controlled preparation of the surfaces to be joined through adhesive bonding, and a method of nondestructive testing that can validate the bond-line without degrading the properties of the adhesive(s) used to hold the two pieces together.

This chapter is about identifying all the bond-line types, technologies, elements, and requirements necessary for applying a technology that can evaluate the integrity of the bond-line, so a confidence level is achieved to replace fasteners with adhesives. Assuming the technology described in the previous chapter is robust and tested, there continues to be concern about whether a chemically bonded structure can sustain its adhesive strength throughout the performance life of the vehicle in which it is used. Adhesives have evolved, but remain suspect over time under the continuous strain, pressure, and extremes of temperature applied during a lifetime of use. Approval at the time of manufacture is one thing, durability over the operating life of the parts bonded together is yet another.

Another concern is fault analysis and failure location identification during nondestructive operational testing. Each successive bond-line is validated as a vehicle is assembled. Using an airplane airframe as a case example, the fuel tanks are pressurized to validate the strength of the bonded assembly. One "sign" that a joint has failed is derived from a very audible ***bang*** sound. When the airplane is assembled, it would be very difficult to identify the location of the point-of-failure if the sound were heard and pressure within the tank dropped. Once depressurization occurs, the pieces previously adhesively bonded close-up and failure point(s) are difficult to locate and repair.[1] And the adhesive bond-line inspection technology size and structure limit access to the interior of the fuel tank to re-evaluate the bond-line. This failure condition created through pressure testing has driven a vigorous search for technologies that complement bond-line inspection technology to advance airframe and vehicle assembly toward fastener reduction or elimination.

No single technology stands alone. While the journey to eliminate fasteners using adhesively bonded structure begins with validating the adhesive bond-line, the development and maturation of other complementary technologies when combined with laser bond-line or other bond-line analysis elevate bond-line analysis and validation to a place where even the most reticent detractors become believers.

The complexity of bond-line inspection requires an approach that has to incorporate multiple technologies. Like any disruptive technology, the need to coalesce many technologies into a solution for adhesively bonding structure is required before fasteners can be replaced by adhesives. As stated in Chapter 8, the challenges for stand-alone adhesively bonded structures are not limited to simple bond-line integrity verification. Technologies need to be adopted that include test, validation, and durability verification.

Close examination of "silver bullet" solutions reveal the planned and accidental development of complementary technologies that are necessary to make the primary (Silver Bullet) technology viable. For example, we often refer to Alexander Gutenberg as the inventor of the printing press. The mechanical operations of printing presses would not have worked if complementary technologies, both planned and accidental, had not occurred.

Wood block printing that used wood blocks containing letters, numbers, and images had existed for centuries. However, the ink would absorb into the wood and degrade its ability to be used for more than a few print iterations. The development of metal type to replace wood had evolved, but was hampered by the drying agents incorporated into the ink applied to the metal blocks' face. The drying agents would corrode the metal and the type would lose its fidelity. New metal and types of ink were developed to facilitate the use of metal to replace wood. Metal replaced wood and could be used for long runs of the same printed material.

The final piece of the printing press technological solution as an evolutionary step from hand reproduced books and newspapers to mass produced paper communication was opportunity and tragedy. The plague arrived and wiped out as much as 90% of Europe. Until the plague, paper was either made of parchment or papyrus. Both were slick and did not absorb ink. They had to be "blotted" to absorb ink. The ink would lie upon the surface so a rapid printing process that stacked rapidly produced replicas out of a mechanical press was not possible. When the plague came, it was indiscriminate,

[1] When a fastener fails, it is very evident because the fastener fractures and comes loose.

killing rich and poor alike. As a result, there existed a dearth of clothing mostly made of cotton. Entrepreneurial spirit met innovation and rag paper was born. Rag paper had the ability to rapidly absorb ink and dry rapidly. Rag paper could be printed upon and then stacked without smearing the ink.

Without the developmental evolution of metal, ink, and paper the printing press mechanical ability to run rapidly and efficiently could not be realized.

The same dilemma exists for laser bond-line technology. Other complementary technologies need to be matured to fully realize the potential for laser bond-line technology to rise to its full potential as the core part of a suite of technologies for acceptance of adhesively bonded structures.

The first step in applying a new technology solution is to identify the requirements necessary to validate its ability to mitigate all the concerns raised by the process change.

10.2 Composite Bonding Definitions

Before beginning the requirements definition for applying a technology to validate two adhesively bonded structures, a short explanation of what is meant by "adhesively bonded structure" is necessary. There has been some confusion about the definition of an adhesively bonded structure, which can be clarified by defining the various methods for joining composite parts together.

10.2.1 Co-curing

Co-curing is a process of curing a composite laminate and simultaneously bonding it to some other uncured material, or to a core material such as balsa, honeycomb, or foam core. With this process, all resins and adhesives are cured during the same process.

10.2.2 Co-bonding

Co-bonding is a process in which two or more parts (elements) are joined together where at least one of the parts is fully cured and at least one is uncured. This process requires careful surface preparation of the previously cured substrate with additional adhesive(s) required at the interface.

10.2.3 Adhesive Bonding (Secondary Bonding)

Adhesive bonding, also called secondary bonding, is the process of joining together, by the process of adhesive bonding, two or more pre-cured composite parts in which the only chemical or thermal reaction occurring is curing of the adhesive itself.

Adhesive bonding requires very careful preparation of each previously cured substrate at the surfaces to be joined together. These are called the bonded surfaces. The adhesive bonding process requires well-designed fixtures to align and clamp the parts together during the adhesive bonding process. While it is a temptation to increase the rate of cure of the adhesive to improve cycle time, re-heating previously cured substrates can be risky and has had catastrophic results, where parts have been scrapped.

The proper amount of adhesive applied for complete squeeze-out for total coverage of the areas to be joined together is essential. Adhesive application must be done without too much over squeeze-out and, when done by hand, is a developmental process derived from a learning process. When applying adhesive to surfaces that must be joined together without the benefit of fasteners, automation may be the only solution that provides the adhesive dispensing uniformity necessary to ensure a robust bond-line.

The automotive industry uses a combination of robots with adhesive dispensers attached that are guided by laser or optics to control the adhesive distribution. Bead inspection technology measures and inspects the bead height, width (volume), and position to avoid leaks, voids and ensures structural integrity once the parts are adhesively bonded together. Dispenser "monitors" also alert users where material is deposited with insufficient volumes by detecting skips, including partial skips.

10.3 Metal Bonding

Metal bonding is the same as secondary bonding (composites), except that with metal bonding the substrates are metal instead of cured composite substrates.

Occasionally, metals are bonded directly to composites using one or more processes. Metals require very stringent surface preparation including application of corrosion inhibiting primer prior to bonding to obtain long-term bond-durability at the metallic interface. As a caution, care must be taken when bonding metal to carbon as galvanic corrosion can occur in the metal substrate.

A significant difference exists between bonded joints in metallic and composite structures. It is possible to design a joint between metals so that the adhesive is always stronger than the metal being joined. This is not always possible with composites because of the significantly higher strength of composite materials. Also the first ply peel failure mode of composites where the relatively low inter-laminar shear or peel strength of the resin system causes failure of the composite before failure of the bond occurs.

10.4 Types of Bond-Lines

If two pieces of composites or metal were simply mated and adhesively bonded, the process of using a bond-line analyzer for quality assurance would be simplified. The variety of bond-lines complicates the bond-line evaluation process and, therefore, also complicates the technology solution for bond-line inspection. **Figure 10.1** shows the primary requirements to ensure a good adhesive bond for a simple single lap joint.

The variety of joint types to be bonded complicates the bond-line evaluation process; and the types of stresses and strains can impact a bond-line evaluation. Tension, compression, shear, cleavage, and peel are all loads that can affect the performance of an adhesively bonded structure. **Figure 10.2** shows the primary loads that can be applied to a structure and, by extension, to the adhesive bond-line during its service life.

Common joint designs include single lap, tapered single lap, single strap lap, double lap, double tapered strap lap, double strap lap, and tapered scarf joints. **Figure 10.3** illustrates why it is more difficult to evaluate the bond-line of certain types of joints because of their complexity.

Adhesive failure results from failure of a bonded joint between the adhesive and the substrate. The cause for adhesive failure includes lack of chemical bonding between the adhesive and the bonding substrate. Failure can also result from poor surface preparation or contamination, or incorrect adhesive selection for the substrate materials. **Figure 10.4** shows a notion example of an adhesive failure.

Cohesive failure is when the failure of an adhesive joint occurs primarily in the adhesive layer. This is an optimum type of failure in an adhesively bonded joint when failure occurs at predicted loads. Failure at lower loads is indicative of poorly cured adhesive or moisture of other contaminants present in the adhesive. **Figure 10.5** shows the complex array of all the potential causes that could contribute to failure of an adhesively bonded structure.

When considering any application of an adhesively bonded structure, it is important to understand how adhesives cure and why it is so difficult to evaluate and validate their integrity. **Figure 10.6** shows the way adhesives cure. Until the joining of all the "cells" is complete, the adhesive properties of the material are not complete. Figure 10.6 shows the cure cycle and how "cells" are formed and join together as the adhesive cures. This condition is further complicated by the time it takes the adhesive to fully cure. The sensitivity to the formation of these "cells" to disruption by impact, pressure, and temperature can affect the integrity of the bond-line.

FIGURE 10.1 Primary requirements for a good adhesive bond.

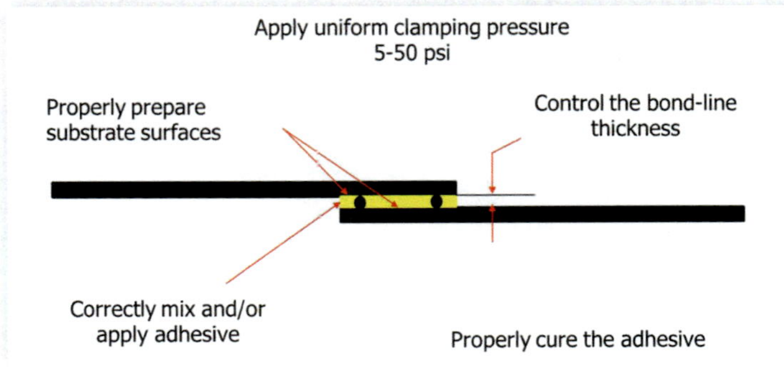

© SAE International/George Bullen

FIGURE 10.2 Loads on adhesive bonded joints.

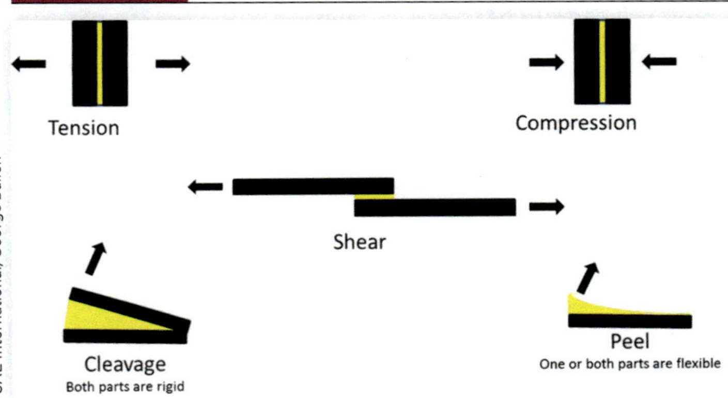

© SAE International/George Bullen

FIGURE 10.3 Complex adhesively bonded joints.

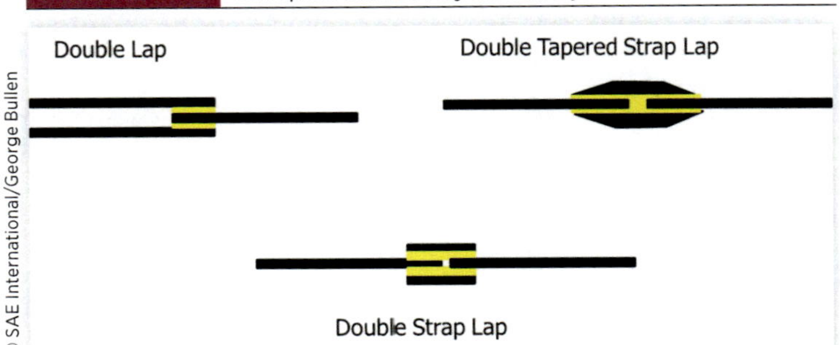

© SAE International/George Bullen

FIGURE 10.4 Adhesive failure.

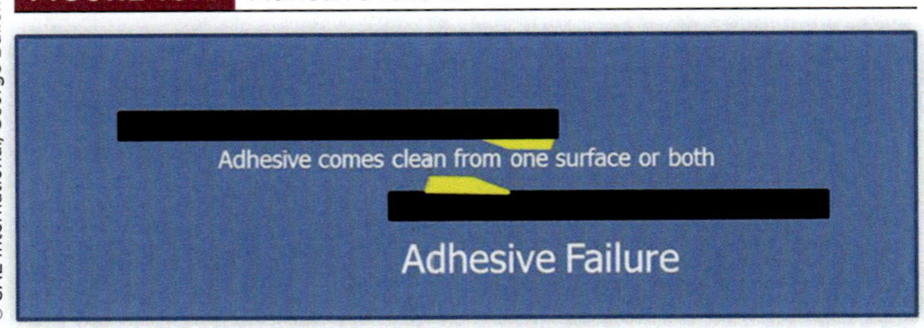

© SAE International/George Bullen

FIGURE 10.5 Adhesive failure fault tree.

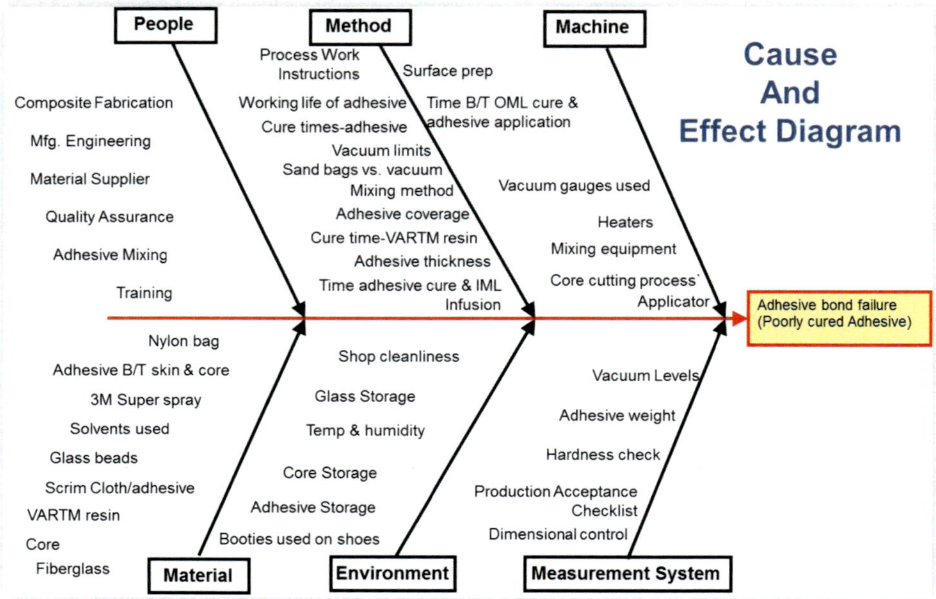

The complexity of adhesively bonded structures has not overcome its advantages. Figure 10.7 shows the wing of the first certified military aircraft (Global Hawk) constructed using an all-bonded primary structure.

FIGURE 10.6 Adhesive cellular joining as adhesives cure.

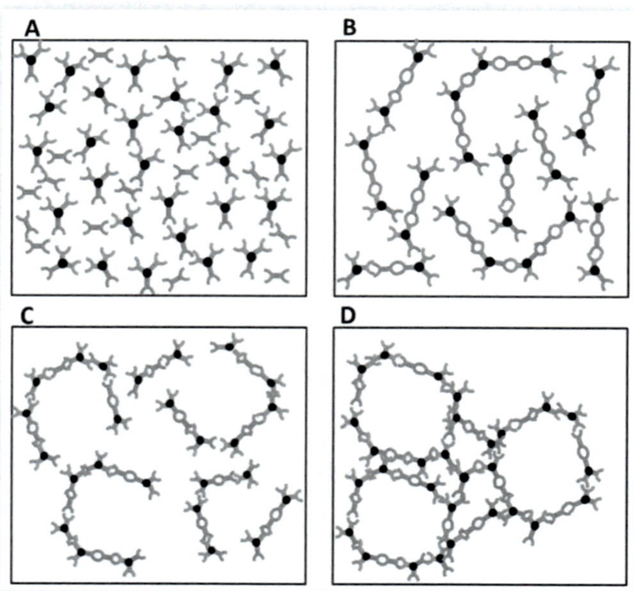

The complexity of validating bond-lines after cure is complicated by the failure of the materials adhesively bonded, even if the adhesive has provided the necessary strength to hold two pieces together. Composites are composed of layers of cured materials that are held together by means of compaction, consolidation, and cure compressed under vacuum.

Even the best adhesively bonded joint can fail because of substrate failure. Inter-laminar fracture in a composite structure usually occurs between the first and second plies next to the bond-line. This type of failure can be common in composite laminates, especially those containing brittle epoxies.

Any nondestructive bond-line inspection technology or group of technologies must incorporate the ability to address all the potential failure modes when two pieces of materials are joined together using adhesives.

FIGURE 10.7 First certified military aircraft (wing) constructed using an all-bonded primary structure.

10.5 Complementary Technologies

What all the complexity means is that there has to be more to evaluating a bond-line to ensure that the adhesively bonded structure "survives" all the operational loads and environments that the product will see over its operational life. Simply validating a bond-line after manufacture is not sufficient.

Apparent from all the complexities previously described regarding adhesively bonded structure is the realization that no single piece of technology will satisfy the criteria to validate a bond-line. There are many technologies that are beginning to fill the gaps that show promise for helping to eventually achieve the goal of an adhesively bonded structure sans fasteners.

Damage localization in composite structures using nonlinear vibration response properties may be one way to detect bond-line failure locations during load testing, and to monitor for bond-line weakness and failure during the operational life of the vehicle.

Subsurface damage and adhesive dis-bonds in composite materials are difficult to detect using visual techniques, and other current inspection methods lack the ability to perform quick, wide-area inspections without the need for reference signatures or baseline measurements. Damage localization in composite structures using nonlinear vibration response properties is a method for detecting and locating subsurface damage in composite materials without historical reference measurements by considering the nonlinear behavior of the material in the vicinity of damage. Nonlinear behavior is identified by comparing frequency response functions measured at different input amplitudes.

The nonlinear behavior of the material is most evident in the areas nearest to the damage. The inspection method demonstrates both analytical and experimental outcomes. First, a finite element model of an adhesively bonded structure is developed using Bernoulli–Euler beam[2] elements to represent each layer of the structure. Bilinear stiffness nonlinearity is simulated to represent dis-bond damage between the top and core layers of the structure. The simulated dis-bond damage is localized by identifying degrees-of-freedom that indicate a significant nonlinear response through a comparison of frequency response functions measured at various input amplitudes.

Next, the method is demonstrated experimentally by identifying dis-bond damage in a fiberglass sandwich panel. A 3-D scanning laser vibrometer is used to measure the forced frequency response of the panel in its damaged state as it is excited at two or more amplitudes of excitation by a piezoelectric actuator. Comparisons of the frequency response functions measured at different input amplitudes show that the subsurface damage introduces nonlinear behavior that resembles bilinear stiffness nonlinearity, and the differences in the frequency response functions are largest in the vicinity of the damage location. Improved localization of the damage is achieved by investigating the response at higher frequencies. This work has application as a nondestructive method for detecting and locating subsurface damage in composite materials and, by using a laser vibrometer for noncontact measurement, it allows for quick, wide-area inspection of composite materials without the need for reference signatures or baseline measurements [10-1].

10.5.1 Case Study: Wide-Area Inspection

Wide-area inspection of aircraft is important in helping to reduce the cost of aircraft maintenance, because it has the capability to reduce inspection times. Wide-area inspection technique(s) for detecting impact damage in composite materials using scanning laser vibrometer measurements and nonlinear forced vibration response characteristics can be a valuable tool. A linear system model is the foundation for either of these types of approaches, however, and as a result, the baseline configuration must be known for comparison. A laser vibrometer for standoff measurements presents a nonlinear damage detection technique that is able to detect and localize damage without measurements of the baseline dynamics of the structure. A 3-D scanning laser vibrometer (Polytec PSV-400-3D) can and has been used to measure the surface velocity of the composite panels as they were excited using a piezoelectric actuator (PCB 712A02). The 3-D laser vibrometer allows the measurement of in-plane and out-of-plane motion giving measurements in three orthogonal directions for flat and curved surfaces, and the scanning capability of the laser vibrometer allows measurements at a defined set of points giving a mapping of the vibration on the surface being measured. This measurement system is ideal for wide-area inspection of aircraft because it allows easy, noncontact acquisition of forced frequency response data over a large, curved measurement surface. Impact damage in carbon fiber sandwich panels can be investigated by measuring frequency response data with the scanning laser vibrometer because of excitation at high- and low-amplitude levels. Excitation at high- and low-amplitude levels allows for subsurface

[2] Besides deflection, the beam equation describes forces and moments and, therefore, can be used to describe stresses.

damage in the panel to be detected without the need for baseline measurements by considering nonlinear characteristics in the forced vibration response. This nonlinear approach is not sensitive to environmental factors in the measurement, such as boundary conditions and temperature fluctuations, since these factors do not change the underlying nonlinear behavior of the damaged composite material.

The goal of reducing the uncertainty and risk associated with scheduled inspection of aircraft is critical to effectively managing the life-cycle costs of aircraft. The two methods previously described, i.e., impact identification and vibrometry-based wide-area inspection, have the potential to reduce the prevalence of labor-intensive inspections of the aircraft that are often unnecessary and/or ineffective. The impact identification algorithm allows for inspections to be focused on areas that are most likely to be damaged, an improvement that could enable frequent inspections of the entire aircraft to be largely replaced with small and more careful inspections of areas that are likely damaged. When wide-area inspections are necessary, the vibrometry-based method presented shows the ability to quickly automate inspections of large aircraft. The combinations of these methods offer the potential to significantly reduce the reliance on inexact and expensive traditional inspection methods. While fully autonomous structural health monitoring (SHM) systems cannot yet replace traditional inspection and maintenance methods, techniques such as those presented offer a step in the right direction by providing maintainers powerful tools to be more effective, especially to determine the integrity where there is an adhesively bonded structure. Utilizing SHM technologies as vital maintenance tools allows the end users, the aircraft maintainers, to gain trust and understanding of the technology. In addition to the concrete benefits that these technologies offer, this increased trust and understanding will help to move one step closer to the ultimate goal of nearly autonomous SHM systems that enable condition-based maintenance and prognostic health management of future aircraft in which the adhesively bonded structure has replaced the fastened structure.

10.6 Bond-Line Deficiency Discovery and Mitigation

The evolution of technologies that can be used in manufacturing and in the field to validate the integrity of bond-lines are significant steps in the right direction for moving toward adhesively bonded structures. “Seeing” the bond-line and assessing its “integrity is one element of a process change from fasteners to adhesively bonded structures. The natural outcome of bond-line inspection is the discovery of nonconforming or deficient bond-lines.

Once discovered, the repair and re-inspection of the deficient bond-line must be performed. As the adhesive bond-line inspection technology has begun to evolve, a rising confidence level has driven manufacturers to look to ways for repairing defects that the inspection technology reveals.

So far the answer that has repeatedly returned as an optimal solution is to add a fastener or fasteners where the bond-line is deficient. This repair strategy seems to be the best alternative so far for defects discovered by a robust combination of manufacturing process adhesive bond-line inspection and operational maintenance and sustainment adhesive bond-line inspection at the depot.

10.7 Summary

Airplanes come in a variety of shapes and sizes to serve specific purposes that provide value and utility to their user. They all have a common core of manufacturing criteria driven by the specific performance demands of their users. Arguably they are one of, if not the most complex machines in the world. Airplanes (fixed and rotary wing) are assembled using a wide variety of both common and exotic materials. Airplanes must provide a safe internal environment for both living passengers and operating systems while performing in extremely harsh environments. For example, at 35,000 ft. (11,000 m) the typical altitude of a commercial jet airplane, the air pressure drops to less than a quarter of its value at sea level, and the outside temperature drops below negative 60°F (negative 51° C). Within 30 min, the same airplane can be on the ground in Phoenix, AZ, USA, and exposed to temperatures in excess of plus 120°F. This cycle of extremes is repeated hundreds if not thousands of times throughout the life of the airplane. And yet air travel is one of the safest forms of transportation in the world. In fact, most crashes occur as the result of pilot error, not mechanical failure. The low rate of mechanical failure is attributable to both the design and rigidly controlled manufacturing process. As a result of the complexity of modern airplanes and the extreme environment in which they operate, draconian and conservative manufacturing processes have evolved to provide the controls necessary for all the millions of assembly operations and pieces of the airplane to be joined together precisely. The manufacturing process is tightly controlled to ensure that the aggregate of all the complex operations are performed to produce a robust product that performs as designed. Overlook a single and seemingly unimportant process involving a minor part and tragedy can happen.

In one of my previous books, *Automated/Mechanized Drilling and Countersinking of Airframes*, I pointed out the fate of Aloha Airlines Flight 243, where a portion of the upper fuselage ripped off at 24,000 ft. (7,300 m) causing explosive decompression. The cause was later determined to be the failure of a single fastener. When building an airplane, everything, even a little one-dollar rivet is important.

The complexity of modern airplanes has also driven subassemblies to be manufactured at tier suppliers where cost and other considerations have resulted in a widely dispersed supply chain. Big companies such as Boeing, Airbus, Northrop, Lockheed, Bell, and others have become assembly centers for subassemblies produced globally by tier suppliers. The exceptions to the global supply chain are core competency materials and manufacturing processes the "primes" want to keep in-house to leverage their use for competitive advantage.

The means of production has also moved out and away from in-house manufacture to a select group of integrators that provide tooling and automation for most of the global airplane manufacturers.

The sensitivity of the manufacturing methods and processes required to ensure a robust bond-line that is durable for the life of the vehicle is absolutely critical. The wide distribution of the supply base for subassemblies complicates the critical control of the required processes necessary for good adhesion.

It should also be noted that adhesives have two time-sensitive elements that complicate their introduction into the manufacturing process. They are work life and cure time.

Work life defines the time from when an adhesive is mixed to the time it must be applied and squeezed-out. The operation is time sensitive and can be extremely limiting. High-strength adhesives can be limited to as little as a 30-min work life.

The cure time once applied and held together can also limit the speed of assembly for parts and assemblies. There are adhesives that can take days to cure; and during the cure cycle certain adhesives are very sensitive to disruption caused by work on the part(s) being cured.

It is an attractive proposition to "glue" an airplane together. It simplifies the process and eliminates the cost, complexity, defects, and weight of a fastened structure. That is why the technologies described in Section Two are being vigorously pursued by Original Equipment Manufacturers. The technologies in development are promising but still lack the fidelity to provide assurance to the operators that adhesives will replace fasteners in both the near-term and mid-term time horizon.

The exception to this statement is uninhabited vehicles, where the vehicles are used over uninhabited areas.

And in road and rail vehicles in which the operational stress, loads, and environments are not as harsh as air vehicles.

Additional Reading

1. Kessler, S. S., Spearing, S., Atalla, M. J., Cesnik, C. E., and Soutis, C. "Damage Detection in Composite Materials Using Frequency Response Methods," Composites Part B: Engineering. 33(1):87–95, 2002.
2. Ghoshal, A., Chattopadhyay, A., Schulz, M. J., Thornburgh, R., and Waldron, K. "Experimental Investigation of Damage Detection in Composite Material Structures Using a Laser Vibrometerand Piezoelectric Actuators," Journal of Intelligent Material Systems and Structures. 14(8):521–537, 2003.
3. Sundaresan, M. J., Ghoshal, A., Li, J., Schulz, M. J., Pai, P. F., and Chung, J. H. "Experimental Damage Detection on a Wing Panel Using Vibration Deflection Shapes," Structural Health Monitoring. 2(3):243–256, 2003.

References

10-1. Underwood, S. S., Meyer, J. J., and Adams, D. E. "Damage Localization in Composite Structures Using Nonlinear Vibration Response Properties," The American Society of Mechanical Engineers, Journal of Vibration and Acoustics. 137(3):031015, 2015.

10-2. Bond, R., Underwood, S., Adams, D. E., and Cummins, J. J. "Structural Health Monitoring Assessment Based Methodologies for Managing Uncertainty in Aircraft Structural Life," October 9, 2014. Available from: http://shm.sagepub.com/content/early/2014/10/09/1475921714553733. Accessed May 1, 2017.

CHAPTER 11

Summary

Physical changes take place continuously, while chemical changes take place discontinuously.

—Max Planck

11.1 Introduction

The technologies described in Section Two of this book highlight an innovative solution to enhance the manufacturer's confidence for bonding structure together without fasteners. When combined with more rigid manufacturing process controls, laser bond-line inspection (LBI) will expand adhesive bonding as a viable consideration for the judicious replacement of fasteners.

It should be noted that bonding together of two pieces of aluminum and especially parts made from composite materials presents more complexities and considerations than simply "making it right" (controlling the process) and verifying the bond-line. The bond created using adhesives is the joining together, by the process of adhesive bonding, two or more pre-cured composite or fabricated aluminum parts, where only chemical or thermal reaction occurs to cure the adhesive itself. It requires careful preparation of each previously cured substrate at the bonding surfaces. This process sans fasteners in a bonded structure becomes the sole means of holding the structure together. Complexities arise from changing from the clamping effect of fastener to holding by adhesives outside the scope of laser bond-line verification.

When using adhesives to bond and hold a part together in areas previously joined by means of fasteners, the clamping effect of fastener is lost. This is particularly relevant when adhesively bonding two pieces of composites together in an assembly. Composite material (also called a composition material or shortened to composite) is a material made from two or more constituent materials with significantly different physical or chemical

FIGURE 11.1 Shows a notional image of a substrate failure.

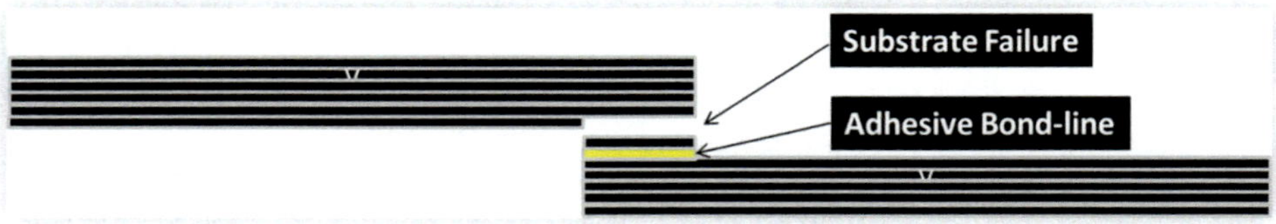

properties that—when combined—produce a material with characteristics different from the individual components. The individual components remain separate and distinct within the finished structure. When making parts, from composites, uncured sheets are layered and pressed together to attain a post-cure thickness. The uncured sheets are rotated, resulting in a cross-grain that provides strength. Parts made from composite materials are also anisotropic. Anisotropy is the property of being directionally dependent, as opposed to isotropy, which implies identical properties in all directions. It can be defined as a difference, when measured along different axes, in the physical or mechanical properties of a material (such as absorbance, refractive index, conductivity, and tensile strength). An example of anisotropy in material is wood, which is easier to split along its grain than against it. When bonding alone is used, the operational stresses along the bond-line can result in the bonded joint failing because of the separation of the layers of material. This substrate failure or inter-laminar fracture in composite structures usually occurs between the first and second plies next to the bond-line. It can be common in composite laminates, especially those with brittle epoxies. **Figure 11.1** is an image of a solid bond-line "defeated" by substrate failure.

If this same bond-line had been enhanced using fasteners, the failure would not have occurred because of the clamping effect of the fastener. **Figure 11.2** shows the same bonded pieces with a fastener through the two work-pieces providing the clamping force necessary to hold the parts together.

Further complications occur in an adhesively bonded structure (AdBS) when dissimilar materials are joined together. When an aluminum part is bonded directly to a part made from composite material, the aluminum requires a very stringent surface preparation process. The surface preparation process includes the application of corrosion, inhibiting primer(s) prior to bonding to obtain and ensure long-term bond-durability at the metallic interface. Bonding metal to carbon can result in galvanic corrosion occurring in the metal substrate.

FIGURE 11.2 A notional image showing fasteners' clamping effect along a bond-line.

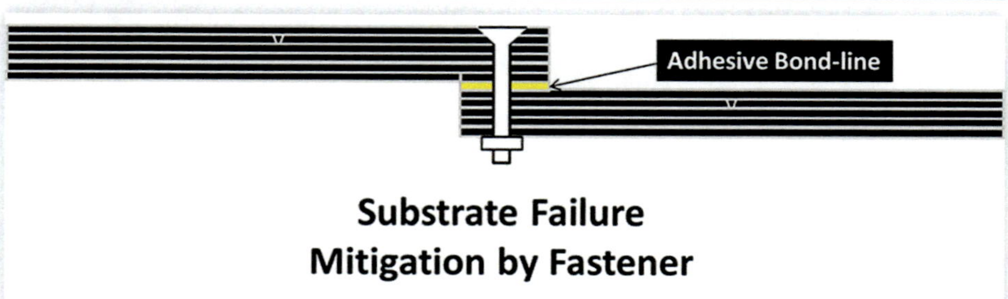

11.2 Adding Complexity

It is important to understand the complexities of joining together an airplane, road, or rail type vehicle assembly using a combination of fasteners and bonded structure. While elimination of fasteners by use of AdBS is attractive to the manufacturers of these types of complex vehicles, the interaction and relationship between joined and fastened structure create complications. Judicious consideration and rigid testing should be applied to determine how these divergent manufacturing processes affect the strength of the joined structure. New design specifications and manufacturing processes must be developed to successfully and economically join parts together into an assembly by means of these two dissimilar methods. Semi-permanent fasteners,[1] when combined with the permanent adhesive-joined work-pieces, complicate the many facets already resident in the highly complex manufacturing world of air, road, and rail transport vehicles. The development of new types of technologies to separate, remove, and replace the pieces of AdBS,[2] without damage to the work-piece or assembly, is equally essential as bond-line inspection technology to fully enable AdBS. The two technologies to assess bond-line quality and provide the ability to remove and replace (R&R) damaged or defective parts will be integral to the migration of AdBS to fulfill its greatest potential.

It may be a dream, or more probably the fantasy envisioned by many manufacturers/assemblers of airplanes and transportation vehicles that the eventuality of 100% glued together vehicles lies just over the horizon. While it is fun to think about, the evolution of a completely AdBS is not practical, nor desirable. The limitations that inhibit a 100% AdBS are resident in economics. Composite materials used for unified structure and the migration of aluminum onto automobile structural components have highlighted the limitations and challenges for incorporation of new materials and processes into a manufacturing model. The lessons that have been learned are transferable to adhesively bonded parts and assemblies.

For example, the potential of composite materials to enable scale-up through unifying multiple pieces or assemblies reaches an economic equilibrium point where savings from unification diminish to zero and begin to add cost to manufacturing, repair, maintenance, safety, and sustainability of the product. Parts that are continuously and progressively unified begin to grow beyond the practical application of secondary operations. In aerospace and shipbuilding, the complexity of secondary operations has resulted in the use of efficiency matrix models for defining the maximum unified structural size that incorporates the intensity, complexity, and amount of secondary operations performed in and around a piece of unified structure. Each assembly design is analyzed and optimized to the best unified structural size and includes efficient access to the structure. Unified structure is size-optimized for access to perform the necessary ancillary operations that must be sequenced to access the structure. In a similar way, a bonded structure will need to develop assembly sequencing and processes to optimize the process and fully realize the potential of bonded structure. The assembly process

[1] Semipermanent fasteners can be removed and replaced, albeit with difficulty. Separation of adhesively bonded structures most certainly will destroy the joined parts.

[2] I have used the acronym AbBS for adhesively bonded structure instead of ABS to avoid confusion with that commonly used acronym for antiskid braking system.

must accommodate the time sensitivity of adhesives to complete the bonding/assembly process before the work life of the adhesive expires.

It is true that the reduction of fastener counts through unified structure in modern airplanes has resulted in significant reductions in manufacturing costs and defects. It is a natural evolution of the desire to further reduce fastener counts by adhesively bonding closeout parts and pieces rather than use blind fasteners to finish an assembly. However, the application of aluminum and composites in air, road, and rail transportation vehicles will continue, but be limited by many factors including repair, maintenance, safety, and sustainability.

11.3 Repair

Repair procedures for AdBS/assemblies need to be developed to include manufacturing and operational applications.

11.3.1 Repair During Manufacture

The technology described in Section Two (when mature) will provide a reliable means of certifying an adhesive bond-line used to join two or more work-pieces. However, the LBI will also unmask defects in the hidden bond-line during manufacture.

Adhesive bonding, especially for joining of primary structures, requires certification of every bonded joint characteristics and conformance to specification. When LBI reveals defects, combining adhesive bonding with mechanical fastening results in a bonded/bolted hybrid joint that can meet certification requirements. A bonded/bolted hybrid joint repair limits cracking in the bond-line so that the structure is still able to carry the specified limit load. In bonded/bolted hybrid joint repair configurations, the mechanical fastener, also called dis-bond stopping feature (DSF), guarantees the structural integrity of the adhesively bonded work-pieces by stopping cracks in the bond-line or reducing the speed of crack propagation to maintain the structural integrity of the assembly [11-1].

The design and development of LBI technology leads to the necessary establishment of a repair process/method for discovered defects. Establishing a joining method to repair the defect in adhesively bonded assemblies with the least impact to cost, schedule, and performance is desirable. Research is now focused on better understanding bond-line cracking behavior in adhesively bonded composite structures and what influence a mechanical repair type fastener will have on its characteristics once installed.

Several investigations have been conducted on bonded/bolted hybrid joint. Scant literature is available to describe research addressing stoppage of crack propagation in the bond-line of a bonded/bolted hybrid joint by means of a mechanical fastener and the resultant internal stress-state after the fastener installation on the bond-line cracking behavior. However, several researchers have used finite element methods to investigate bonded/bolted hybrid joints in terms of damage mechanisms. Kelly [11-2], [11-3] used a single lap joint to show that, in most cases, cracking occurs in the bond-line because of the peel stress at the overlap ends. Fu and Mallick [11-4] assert that the mechanical fastener yields a reduction of the peak stress in a bonded/bolted hybrid joint. Crack initiation and propagation depends on the load transfer level as Potter shows in his research [11-5].

Investigation has been performed for the short- and long-term bond-line cracking behavior in a bonded/bolted CLS hybrid repaired joint. The evaluated CLS specimen consisted of two adherents made of the composite material type IM7-8552, a mechanical fastener described by the standard ASNA 2042 (bolt)/ ASNA 2041 (collar), and a bond-line.

The short-term behavior was investigated using finite element simulation. A simulation model was developed and validated considering the adherent strain, the force–displacement curve, the crack front shape, and the crack length–load curve. The validation was carried out through comparison of the simulation results with those of the experiments. Quasi-static loading of the CLS specimen was simulated, and bond-line cracking was implemented using a cohesive zone model.

Cracking stopped in the front of DSF because of the interaction between the fastener form fit and the preload, resulting from the fastener installation. The bond-line peel behavior, which is suppressed by the fastener form fit and the preload traction, has been revealed to be the bond-line cracking carrying factor.

The cracking stopper potential of the mechanical fastener (DSF) depends on the used adhesive. The number of mechanical fasteners needed in a bonded/bolted hybrid joint can be considerably reduced using a soft adhesive. Accordingly, soft adhesives seem to be more appropriate in the bonded/bolted hybrid joining concept than brittle ones and present a high weight reduction potential. These findings have been confirmed by investigation of the long-term bond-line cracking, which was conducted experimentally by means of fatigue testing [11-2].

Further work needs to be done to establish a database and provide standard repair procedures for defects in the bond-line discovered by technology such as LBI.

11.3.2 Repair Operational Damage

Air, road, and rail vehicles are some of the most complex mass produced machines in the world. What make them even more complex are their integration, interaction, and operation with humans.[3] Even autonomous "drones" or uninhabited vehicles must integrate with the unpredictability of the great variety of human actions that are often illogical and erratic. When at Rohr Industries in the 70s, I was an engineer on the team for the design, development, and manufacture of the Bay Area Rapid Transit (BART) cars. We designed and manufactured an all-composite (fiber glass) unified structural operator cab that was fastened onto the aluminum main rail car section by means of fasteners. **Figure 11.3** shows a BART car with its unified structure all fiberglass composite operator cab section attached to the aluminum passenger section of the rail cars. I took this picture of the test car(s) just before it was to perform an impact test. Rohr Industries had the capability and adhesives were available that would have made it possible to adhesively bond the crew cab to the

FIGURE 11.3 Image of BART car with fiberglass composite crew cab and Lexan windshield.

© SAE International/George Bullen

[3] Most of us have seen the car in front of us turn on the left blinker and then suddenly turn right.

aluminum passenger section. However, the United States Department of Transportation (USDOT) and California Department of Transportation (Cal DOT) made it a requirement and written specification directing that the crew cab survive a 70-mph direct impact from a washing machine suspended from an overpass. The USDOT and, by extension, Cal DOT had experience with this type of "event" in Chicago and other cities where rapid transit had been in use for decades. Apparently, humans amuse themselves by dangling everything from cantaloupes to washing machines from overpasses and the entrance of tunnels to witness the resultant explosion and debris from impacting the rapid rail cars. Many operators had been injured and several killed as a result of the impact. This design consideration drove us to fasteners rather than adhesives. Fasteners would facilitate a more rapid removal and replacement of the crew cab in the event of one of the larger "human-entertainment" impacts. The driver's window was also hardened using 2-in.-thick ballistic Lexan, which could withstand the washing machine impact and protect the driver from occasional bullets directed at the train. Fastening, rather than bonding, facilitated the rapid replacement of the windshield if damaged.

Many parts of an assembly are specifically designed as an attachment to the main assembly because of their proximity to a harsh operating environment. Permanent fasteners are installed to attach these parts to the vehicle with the understanding that during the operational life of the vehicle, the parts must be removed for repair or replaced.[4] Two examples are the undercar environmental ducting for the BART cars and airplane parts in which the engineers have designated parts as remove and replace (R&R) or sometimes remove, repair, and replace (RR&R). The BART cars have undercar environmental ducting that is composed of an adhesively bonded inner shell, covered with insulation and encapsulated in an adhesively bonded four-piece outer shell. Each 3–4-ft section of the ducting snakes its way underneath the BART cars where there exists a harsh operating environment. Each section is fastened together with the mating ducts interlocking lap joint. The harsh operating environment necessitates a 2–5-year rotation schedule for duct repair or replacement. The ducts are hung to the under-car body section with fasteners so they can be removed and replaced without destroying the ducts.

For aircraft, R&R parts are generally designated in a similar fashion. When parts are located near harsh operating environments, wear and abuse necessitate the removal and replacement of the parts. "Permanent" fasteners are drilled out, the part is removed, and a replacement part is installed with new "permanent" fasteners.

Composites and aluminum materials have been added to air, rail, and transportation vehicles for many reasons, including weight reduction. Aluminum and composite materials have also enabled manufacturing cost reductions through their ability to be "scaled-up" using unified structure and adhesive bonding. While these materials have added value to the fuel efficiency, performance, and manufacturing cost and defect reduction of vehicles, they have complicated the repair process.

In the book, *Aluminum Auto-Body Joining*, data are presented that show the economic impact of fusing aluminum in a multi-material lightweight vehicle via studies of joining dissimilar materials, joining methods, and performance of the

---

[4] Permanent fasteners are installed due to engineering design requirements, performance requirements, or because the part life may equal the vehicle life.

joined materials [11-6]. Repair costs have increased where composite materials and aluminum have replaced steel in road vehicles and composite materials have replaced aluminum on airplanes. Auto-body shops equipped to repair steel are resistant to invest and incorporate the necessary clean rooms and treatment centers that aluminum and composite vehicles require. Repair shops replace the damaged part or area, or more often recommend scrapping the vehicle. As a result, the operating cost in fuel consumption is reduced, but the insurance cost increases to offset the higher repair costs.

Another aspect of repair, previously not impactful, prior to unified structure and AdBS is access to assess corrosion and backside damage.

In the case of aluminum, where an aluminum part is adhesively bonded to another aluminum part, or where an aluminum part is adhesively bonded to a dissimilar material such as steel or composites, corrosion can defeat an otherwise robust bond-line. The corrosion often occurs behind the observable outside surface and within the critical load bearing surfaces of the adhesively joined parts. The unobserved corrosion potential will necessitate a second type of innovative technology to "see through" the skin to determine if there is corrosion underneath the observably surface that could cause failure.

An example of the hidden and unpredictability of backside corrosion is resident in the Northrop Grumman E-8 Joint Surveillance Target Attack Radar System (Joint STARS). A United States Air Force Airborne ground surveillance, battle management, and command and control aircraft. It tracks ground vehicles and some aircraft, collects imagery, and relays tactical pictures to ground and air theater commanders. The idea was to mount sophisticated electronic equipment onto Boeing (BA) 707 retired legacy airframes to economically remanufacture them for their mission. The thought was by acquiring retired legacy airframes, the cost would be reduced through elimination of a newly designed and built airplane.

Although the best-of-the-best 707 airframes were selected, the surprise came when corrosion was discovered under the skins, between the skin, and the substructure in areas hidden from view and inaccessible. Without a means to see through the skin to find areas where the corrosion existed, the fasteners had to be taken out and the skins removed. Had the skins and substructure been composed of AdBS, this process would not have been possible. Repair or replacement of parts made from aluminum that are adhesively bonded requires the damaged part to be completely removed all the way to its first available fastened joint, or cut and welded.

Where hidden backside corrosion exists with aluminum, hidden backside damage assessment exists to an even greater extent with composites. A composite part can have a minute visually observable impact point that has catastrophic backside damage. **Figure 11.4** shows a small observable impact on composites with backside damage.

Undetected or ignored, the propagation of backside damage could result in part failure.

FIGURE 11.4 A small observable impact on composites with resultant backside damage.

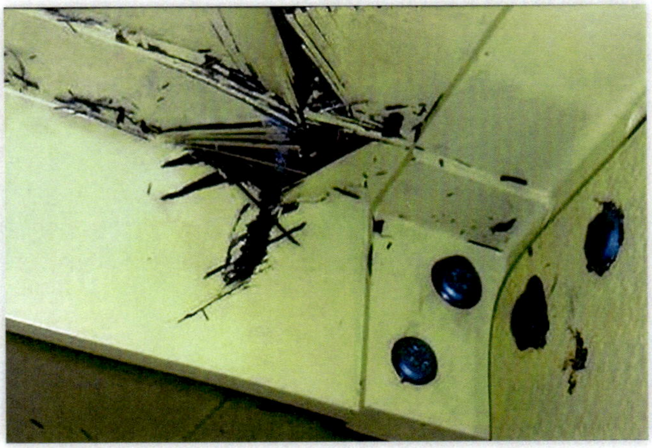

© SAE International/George Bullen

FIGURE 11.5 A repaired composite part on an airplane.

© SAE International/George Bullen

Finding and repairing large backside damage is complex and expensive in unified or AdBS. Figure 11.5 shows an example of composite repair where the AdBS necessitated cutting and replacing a damaged part.

New repair processes, repair facility investment, and training will have to grow hand in hand as new materials and bonded structures make their way into new air, road, and rail vehicles.

11.4 Maintenance

The complexity of operating systems for mass produced air, road, and rail human transport systems requires access to their interior for sustaining maintenance. Pipes, tubes, ducts, wires, and cables internally connected to pumps, monitors, sensors, and systems that wind their way through the length of the vehicle and are necessary to bring life to the vehicle-shell and sustain its operation. The necessity of passing through the interior structure of the vehicle also necessitates breaks in a tube or wire in which fittings have to be installed for continuity or connected to the pumps, monitors, sensors, or systems. Access to these operating systems and their "life-lines" to check, calibrate, and sustain their operation is accomplished through the current design of access points using permanent or removable-type fasteners. By combining pieces through bonding, redesign and reconsideration of access to these critical components will have to be evaluated. While it is nice to think that the hidden and critical system components of air, road, and rail vehicles will be operable at an acceptable performance level over the life of the vehicle, it is not true. Access for repair is a part of the total life-cycle cost of the operation of a vehicle. Encapsulating the pipes, tubes, ducts, wires, and cables internally connected to pumps, monitors, sensors, and systems leads to dramatic increases in maintenance cost.

11.5 Safety

Composites, the wonder material with lightweight, high strength-to-weight ratio and stiffness properties have come a long way in replacing conventional materials, including metals and wood products. The composite technology of a matrix reinforced with manmade fibers such as glass, Kevlar, and carbon after meeting the challenges of aerospace sector have cascaded down for catering to industrial and domestic applications such as road and rail transport vehicles.

The sheen of the wonder material began to wear off 6 years ago, when a B-2 bomber's engines caught fire during what the Air Force called a "routine" engine start.

The fire was one of three that the Air Force studied "to identify new tools and techniques that will allow firefighters to more efficiently cut, penetrate, and extinguish burning aircraft made of composite materials."

Such fires are challenging because hidden interior fires are difficult to extinguish, composites smolder and reignite, and fuselage penetration is virtually impossible with an axe and difficult with a K-12 (fire-fighting saw). The same difficulties would complicate the extraction of victims of crashes where composites were used as the material to construct the bodies of transportation vehicles.

Also, the smoldering composite vehicle emits toxic smoke that may be harmful to rescue workers, witnesses, and bystanders. The physical and chemical properties of the particulate fraction of carbon-graphite/epoxy advanced composite material (cgeACM) smoke were measured to address concerns regarding potential health hazards posed by the release of fibers during pyrolysis of this material. Filter, low-pressure cascade impactor, and electrostatic precipitator samples were collected from cgeACM smoke in which the aerosol concentration ranged from 0.20 to 5.39 g/m^3. Fibers were found in the smoke among individual, spherical, or nearly spherical particles and chain aggregates. The fibers had a mean count diameter of 0.54 micron and an average length of 2.84 microns. However, fibers accounted for approximately 0.3% of the particles counted. The smoke aerosols (including fibers) had mass median aerodynamic diameters (MMAD) ranging from 1.4 to 1.9 microns with standard geometric deviations ranging from 1.6 to 1.8, and hence more than 88% of the particles were in the thoracic size range (MMAD< or = 4.0 microns). All particles were composed primarily of carbon, silicon, sulfur, and oxygen with traces of other metals. By comparison, fibers were composed almost exclusively of silicon (Kimmel 2012) [11-7].

Asiana Airlines Flight 214 wreckage was strewn along Runway 28L at San Francisco International Airport after it crashed on July 6, 2013. The tail section had been torn from the wide-body BA 777 during the ill-fated landing. One of the huge engines was ripped from the aircraft's wings. The fire eventually incinerated the aircraft's core with such intensity that it burned through much of the roof. A gaping gash atop the fuselage revealed a fire-gutted interior from the conflagration that engulfed the airliner soon after it came to rest next to the tarmac.

As a result of these and other airplane crashes where composite materials were used, the past few years have seen the emergence of research directives to address the operational safety challenges of operating a vehicle made from composite material. They include developing a new flammability test method for composites, composite pre-pregs that display fire resistance, flame retardant adhesive technology for interior applications, modified cyanate ester for air duct applications, and fire-resistant nano-coatings for foam and fabric using renewable and/or environmentally benign materials. In addition, emergency response teams must be trained and equipped with the new tools to enable the rescue of individuals trapped in damaged vehicles.

On a positive note, we have been amazed to see passengers walk away from crumpled, singed wrecks like the Asiana plane or Air France Flight 358, which turned into a fireball after overshooting a Toronto runway in 2005, or another BA 777 that crashed on landing at London's Heathrow airport 5 years ago, but saw no fatalities. The survival of the passengers was partially attributed to the strength of the airframe and other structural components made of composite materials.

The lesson is that the more we unify structure and bond pieces together and then place humans into the resultant vehicles, the more difficult it becomes to extricate them when they become crash victims. And the fires and fumes that result from crashes can have a chilling effect on survivors who think they have survived a crash because of the tough material and unified structure, only to find themselves trapped by the same resilient material and manufacturing process that saved them from injury.

11.6 Summary

The previously listed cautions and considerations may seem daunting and discourage the pursuit of AdBS. However, there are examples that incentivize manufacturers willing to invest in its application. One notable example was validation of cost savings during the Rapid Airframe Production Integration Demonstration (RAPID) project. The aerospace industry is undergoing significant changes in the way Department of Defense (DoD) aircraft programs are awarded. The shift is moving away from the award of large multiyear programs such as the F-35 to smaller incremental contracts that may involve the initial fabrication of only 1–5 vehicles. The DoD also requires that the programs be completed at a fraction of the cost and schedule of what is typical. Northrop Grumman Corporation (supported by Spirit Aero Systems for detailed part fabrication) used RAPID to take a completely fresh look at how composite aircraft are designed and built using adhesive bonded structure as one of the primary components for assembly.

Under the RAPID program, funded by the Defense Wide Manufacturing Science and Technology program of OSD and managed by the Materials and Manufacturing Directorate of AFRL, Northrop Grumman completed the assembly of a complex, bonded composite, center fuselage structure. The 18-month effort involved conceptual redesign, fabrication, and assembly of a center fuselage section modeled after Northrop Grumman's X47-B. The selected fuselage section is one of the most challenging areas of an aircraft for application of composite structures. The complexity of the bonded structure can be seen in the incorporation of an integrated composite duct and very highly loaded "cat" and "trap"[5] and nose landing gear metallic bulkheads.

RAPID demonstrated methods to fabricate and assemble airframes at lower cost. The project team built extensively on the work performed under AFRL's Composites Affordability Initiative. Application of out-of-autoclave cure composite materials, 3-D woven preforms, and bonded joints as well as other individual technologies used throughout the industry were combined with novel fabrication methods that demonstrated "faster, better, and cheaper" manufacturing processes and methods. This resulted in a composite airframe that "snaps" together with very little fabrication or assembly tooling compared to conventional composite structures.

Preliminary estimates predict a reduction in nonrecurring tooling costs in excess of 50% and recurring assembly cost reductions of approximately 15%. These savings do not take into account the performance benefits associated with the removal of thousands of airstream fasteners. The concept is also "change friendly," enabling modifications very late in the design cycle that are typically very expensive and extremely disruptive to programs. If successfully transitioned on future programs, costly assembly jigs, autoclaves, and most of the traditional composite cure tools may become a thing of the past.

Lessons learned from the RAPID program are directly transferable to commercial aviation, road, and rail vehicles. The bonded assembly reduced fastener counts by 68% and

[5] The terms "cat" and "trap" are used in United States Navy aviation jargon to describe a catapult assisted takeoff (Cat) and a landing hook that extends from the airplane and catches a cable located on an aircraft carrier to help decelerate the airplane commonly referred to as trapping or "trap."

enabled midcourse design changes with minimal cost or schedule impact. Figure 11.6 shows the completed RAPID fuselage structure.

While challenges exist to realize the dream of an air, road, or rail vehicle that is mostly "glued" together, and challenges exist to manufacturing processes and repair/maintenance of an AdBS, the benefits may well be worth the challenges. Laser bond-line assessment and inspection technology and robust manufacturing process controls will help enable the journey to a highbred vehicle combining the best of unified structure, adhesive bonding, and fasteners.

One additional benefit of bonded composite structure is its ability to "accept" embedded health monitoring sensors and antennae. Health monitoring sensors and antennae can more easily be incorporated into composite materials if fasteners are not present. The metal component of a fastener can disrupt signals and complicate the design of these advanced systems. They also complicate the design and manufacture of a fastened-together airframe by having to strategically place sensors and antennae so as not to be damaged by the fastener installation process. By bonding structure together, an antennae or sensor can extend to the edge of a part without consideration for weaving in and around holes and fasteners.

FIGURE 11.6 Image of the completed RAPID fuselage structure.

References

11-1. Baha II, S. "Cracking Stopping in the Bond-Line of Adhesively Bonded Composite Adherents by Means of a Mechanical Fastener: Numerical and Experimental Investigation," SAE International Journal of Materials and Manufacturing. 9(1), 2016. doi: 10.4271/2015-01-2611.

11-2. Kelly, G. "Quasi-Static Strength and Fatigue Life of Hybrid (Bonded/Bolted) Composite Single-Lap Joints," Composite Structures. 72(1):119–129, 2006.

11-3. Kelly, G. "Load Transfer in Hybrid (Bonded/Bolted) Composite Single-Lap Joints," Composite Structures. 69(1):35–43, 2005.

11-4. Fu, M., and Mallick, P. K. "Fatigue of Hybrid (Adhesive/Bolted) Joints in SRIM," International Journal of Adhesion and Adhesives. 21:145–159, 2001.

11-5. Lee, E. U. "Effect of Load Transfer on Fatigue of Mechanical Fastened Metallic Joints." In Potter, J. M. (ed.), ASTM STP 927, pp. 95–117. American Society for Testing and Materials, Philadelphia, 1986.

11-6. Bullen, G. N. Aluminum Auto-Body Joining. SAE International. Warrendale, PA. Product Code of PT-173, ISBN: 978-0-7680-8252-4, November 6, 2015, 2009.

11-7. Kimmel, E. C., and Courson, D. L. Characterization of Particulate Matter in Carbon-Graphite/Epoxy Advanced Composite Material Smoke. Geo-Centers Inc., Naval Health Research Center Detachment (Toxicology), U.S. National Library of Medicine, National Institutes of Health. Available from: PubMed.gov.

CHAPTER 12

Closing Words

Final thoughts are so, you know, final. Let's call them closing words.

—**Craig Armstrong**

12.1 Technology Evolution

This book has been written to identify the primary challenges associated with fabrication and assembly of air, road, and rail vehicles. And to identify the technologies that can mitigate these challenges to enhance quality, leverage material properties for best outcomes, and reduce cost. New materials such as high-strength carbon fiber (HSCF) composites have migrated onto air and road vehicles and offer value to the manufacturer and user that has yet to be fully realized. Composite material continues to challenge manufacturers who have deep roots in metal as the primary material for part and product manufacture. In the foreseeable future, fasteners will continue to be the primary means to join all the disparate vehicle parts made from aluminum or composite materials into a cohesive and functioning machine. **Figure 12.1** shows a composite airplane subassembly joined together by means of fasteners.

HSCF material and manufacturing complexities attributed to them have become better understood, revealing that the transfer of metal-specific technology such as gauges is inadequate to ensure that fastener hole-bores and countersinks conform to specification. New technologies are emerging to automatically validate these two critical components for fastener strength and fidelity. **Figure 12.2** shows a defect discovered in a hole-bore in composites by a noncontact imaging ring laser that was passed and escaped detection by the traditional-type gauges meant for use on metal used today (image courtesy of

FIGURE 12.1 Composite airplane subassembly joined together by means of fasteners.

© United Sciences

FIGURE 12.2 Escape defect found by new noncontact imaging ring laser technology.

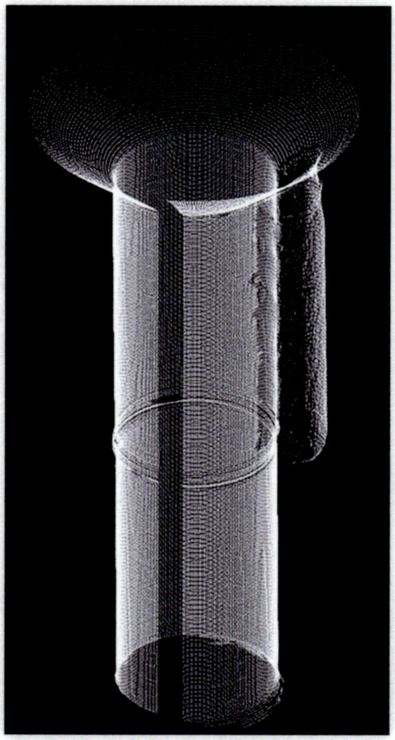

© United Sciences

United Sciences). The new technologies are also necessary to fully leverage automated systems and remove humans from the fastening process. Automation has spread across the drill and countersink manufacturing landscape, but widespread automated installation of fasteners has been defeated by the interim operation for gauging hole-bores and countersinks manually. **Figure 12.3** is an example of one type of drilling system for airframe assembly.

Installation between the drill and countersink operation, and the fastener installation process of an automated gauging system capable of evaluating both composites and metals is needed. The technology resident in ring laser noncontact gauges replete with diagnostic software binds the current disparate operations together into a fluid single automated process. The fidelity of the software also enables connectivity through the digital thread across the three operations (drill/countersink, gauge, and fill) to facilitate a coordinate process for best outcomes. When linked into the complete manufacturing digital tapestry, the gauging system provides cross-functional diagnostics to enhance manufacturing best fit at assembly.

As HSCF composite materials and the complexities of fastening them together have been recognized, efforts have been made to unify structure to eliminate fasteners. Unified structure offers many other benefits besides fastener reduction, but reducing fastener count provides real and tangible benefits to vehicle manufacturers who use HSCF composite materials. **Figure 12.4** shows the NASA composite crew module (CCM) built using unified structure design principles.

FIGURE 12.3 Drilling system for airframe assembly.

© Boeing

FIGURE 12.4 Composite crew module (CCM).

© NASA

FIGURE 12.5 Bolted bracket on CCM.

© NASA

The manufacture and assembly of unified structural vehicles still require fasteners. Figure 12.5 shows bolted metal fittings attached to the CCM (image courtesy of NASA). Fasteners will in all likelihood never be completely eliminated any more than automation will replace all human activity in the assembly operations for air and road vehicles. Concerted ongoing fastener reduction efforts for current and future air and road vehicles will result in fewer and fewer fasteners to assemble these complex machines.

Advancements in manufacturing technology and materials will enable the construction of an increasingly larger unified structure for application on air, space, and road vehicles. The cryogenic tank being fabricated at Boeing Airplane Company for NASA shown in Figure 12.6 incorporates manufacturing and material technologies that are directly transferable to airplanes and road vehicles. The types of technologies illustrated in Figure 12.6 demonstrate the viability of scaling down or up to meet the size needs and requirements for a technology transfer.

Further enabling fastener count reduction are the attributes inherent in HSFC composite materials to provide surfaces that can be glued together. Size and shape restrictions, fabrication and assembly sequencing, and access considerations are some limiting factors for upsizing and incorporating unitization into a part design. Post-cure-process joining methods now incorporate fasteners even if the parts are adhesively bonded. The natural evolution of the adhesion process is to consider the bond-line and the ability of adhesives to stand-alone as a method for holding the pieces of an assembly together.

The hurdle for taking away the fastener "security blanket" is the ability to analyze the bond-line after cure to determine if all criteria have been met for a strong bond to meet engineering requirements. Absent the ability of the assembly manufacturer to

FIGURE 12.6 Composite cryogenic tank during fabrication using automated manufacturing technology.

© NASA

"see" through bonded material and into the bond-line, several approaches have been suggested to remove the fear associated with a dis-bond. Dis-bonds act like fastener failures. When a fastener fails, the stress is transferred to the adjacent fastener, increasing the load on those fasters. Depending on the location of the fastener at failure and many other factors, one or both of the adjacent fasteners may fail under the increased stress. When one or both of the adjacent fasteners fail, the load is transferred to their adjacent fasteners. This process can continue opening up and separating the pieces that were joined and held together by the fastener. The common term for this is a "zipper effect" as the skin or pieces separate like the opening of a zipper.

The fear when bonding structure without the means to evaluate bond-line integrity is that this zipper effect will happen when the robust and disciplined process for joining pieces together with adhesives breaks down. There are advocates that promote a trusting process where the manufacturing process becomes so robust, controlled, and understood that the advanced computer models, advanced analytics, Bayesian process control, and vigorous in-process monitoring would provide sufficient assurance of bond-line integrity.

FIGURE 12.7 Failure of an adhesively bonded structure.

© NASA

They have yet to convince the buyers of air and road vehicles that this "trust" process is valid for performance and safety critical assemblies. Stuff happens as they say. Figure 12.7 is a picture of a section cut from adhesively bonded structure in which presumably the manufacturing process had been vigorously followed. The Plexus adhesive had failed to cure, resulting in an aesthetically beautiful part without any outward indication of its dis-bond. The dis-bond would have caused catastrophic failure during flight. But for the smell when the technicians opened the box, the defective bond-line would have gone undiscovered until the part broke into pieces during flight. The effect of the dis-bond discovery was initiation of a forensic engineering analytical study that delayed production for months and cost a lot of money. Out of the hundreds of potential causes, it was discovered that the cause was a mechanic's simple changing of a sequence that to him seemed inconsequential. Fortunately, the vehicle was an uninhabited prototype that would not have cost a life if the defect had not been discovered, but the point is well taken that the adhesive bond-line philosophy should be trust, but verify. Until the bond-line can be verified, use fasteners. It should be noted that since this was an uninhabited prototype, the decision had been to go without fasteners through the bond-line.

No method of nondestructive testing or inspection has satisfactorily demonstrated the capability to quantify the long-term strength and durability of bonded joints. Proof testing of each critical bonded joint on every new aircraft (application of the limit design load to verify that the bond does not fail) would be prohibitively expensive. Historically, mechanical fasteners through the bond-line thickness have been the design feature used to provide safety margins for the unknown.

There are advantages for joints without fasteners that go beyond eliminating fastener hardware and the efforts associated with their installation. Component thickness can be reduced, more efficient load paths are possible, and thousands of stress concentrations are removed. As previously mentioned, their removal can lead to significant additional savings in weight and greater airframe structure optimization.

A technology with software to "see" through material and into the bond-line while possessing the evaluation capability to analyze, assess, and validate or reject areas of a bond-line is being vigorously pursued by companies such as Boeing, Lockheed, and Northrop Grumman. Absent the evolution and transfer to production of bond-line assessment technology, adhesively bonded structures will continue to use fasteners to provide assurance and insurance to users and manufacturers. Until the technology matures, adhesively bonded structure proponents will be stuck with the dilemma of having to break (destroy) the part to prove the bond-line meets specifications while realizing that when they break the part to prove the bond-line, it is no longer useable. They will have to continue to use fasteners.

When the technology for bond-line evaluation matures and becomes a common process tool for bond-line assessment and validation, fasteners will continue to be used as fixes where the technology determines the bond-line has weaknesses.

When an area of a bond-line is determined by a nondestructive method to have areas of nonconformance (weak bond-line), the fastener will re-emerge as the savior of the bond-line. Areas of deficiency will be shored-up using this age-old joining method.

Additional Reading

1. Baja II, S. "Cracking Stopping in the Bond-line of Adhesively Bonded Composite Adherents by Means of a Mechanical Fastener: Numerical and Experimental Investigation," SAE International Journal of Materials and Manufacturing. 9(1):90–96, 2016. doi: 10.4271/2015-01-2611.
2. Kelly, G. "Quasi-Static Strength and Fatigue Life of Hybrid (Bonded/Bolted) Composite Single-Lap Joints," Composite Structures. 72(1):119–129, 2006.
3. Kelly, G. "Load Transfer in Hybrid (Bonded/Bolted) Composite Single-Lap Joints," Composite Structures. 69(1):35–43, 2005.
4. Fu, M. and Mallick, P. K. "Fatigue of Hybrid (Adhesive/Bolted) Joints in SRIM," International Journal of Adhesion and Adhesives. 21:145–159, 2001.
5. Lee, E. U. "Effect of Load Transfer on Fatigue of Mechanical Fastened Metallic Joints." In Potter, J. M. (ed.), ASTM STP 927, pp. 95–117. American Society for Testing and Materials, Philadelphia, 1986.

about the author

George N. Bullen is an internationally recognized expert and consultant to industry for the manufacture of fixed and rotary wing air vehicles, rockets, missiles, and space vehicles. He is the founder of the International Aerospace Automation Consortium and co-founder of the International Economics of Composites Symposium. Mr. Bullen is also a Technical Fellow in Advanced Automation and Global Manufacturing at Northrop Grumman Corporation. Being a fellow of the Society of Manufacturing Engineers and Certified in Production and Inventory Control Management (CPIM), Mr. G. Bullen has an MBA from Loyola Marymount University and a BSMG degree from Pepperdine University. His expertise includes inhabited and uninhabited aerial vehicles, space vehicle design and manufacture, laser weapon system design and manufacture, and lean processes and applications. He has been awarded 16 U.S. and international patents for technology innovations related to manufacturing, mechanization, robotics, robotics control software, and nuclear testing/quality devices, which are the basis for all current automated systems for the assembly of airframes in the United States and Europe. He maintains membership on the academic boards of major universities and is a member of the steering committees of professional societies. He is widely published in magazines, conference proceedings, and peer-review journals. He received SAE International's Forest McFarland Award in 2014 and the American Institute of Aeronautics and Astronautics Design Engineering Award in 2000 for significant advances in aerospace engineering. Mr. Bullen is the author of five books (four published by SAE International) covering a wide array of subjects including innovative technologies, economics, composites, robotics, space launch vehicles, and autobody joining. His next book ***The Future of the Airplane Factory: Digitally Optimized Intelligent Airplane Assembly*** is due to be published in early 2018.